건축물 속에
숨겨진
디자인 건축
하랑
도서출판

건축물 속에 숨겨진 디자인 건축

발행일 : 2025년 11월 28일
출판사 : 하랑출판
주　소 : 서울시 중구 퇴계로28길 8
전　화 : 02. 2263. 3337

CONTENTS

Voysey Charles Francis Annesley

After gaining experience in the London offices of J. P. Seddon and George Devey, Charles Francis Annesley Voysey (1857–1941) set up on his own in 1882. Although he was much influenced by the Arts and Crafts movement, he followed his own, less backward-looking style, which was orientated more towards the functional than the ornamental. Apart from building houses, he was also interested in their interiors, the simplicity of which was further highlighted by his preference for unpainted wood surfaces. He desire was for everything, right down to the toothbrushes (although they were something Voysey does not appear to have designed), to be chosen by the architect. He also designed furniture and textiles for Alexander Morton & Co. and wallpapers for Essex & Co. His extensive oeuvre consists mainly of country houses, often with slate roofs, rough plasterwork, ribbon windows and high saddle and hipped roofs. The section of the middle classes that prospered with industrialization wanted the kind of comfortable home that previously had been the preserve of the aristocracy; they were now the clientele, and Voysey provided the setting for their lifestyle. Perrycroft House (1893) was realized in Malvern, England, for J. W. Wilson and Greyfriars House in Surrey, England, in 1896. Annesley Lodge in Hampstead, London (1895–1897) was built for Voysey's father. Although the L-shaped house has the same angled wall protrusions as Perrycroft House, it lacks the latter's details, such as the tower and round windows. The white-painted pebbledash Broadleys House on Lake Windermere, England (1898–1899) was built for Arthur Currer Briggs, who brought weekend guests over the lake by boat. On the west side, three large bay windows with horizontally latticed glazing offer panoramic views, while relieving the 60-centimeter-thick walls. At the same time, Voysey designed Moorcrag for J. W. Buckley, the clear, almost rectangular layout of which is covered by an elaborate roof landscape. In 1899, he built The Orchard for himself, a country house near a railway station that incorporated all of his developments.

"There are certain qualities that may be regarded as essential to all classes of home... repose, cheerfulness, simplicity, breadth, warmth, quietness in storm, economy of up-keep, evidence of protection, harmony with surroundings, absence of dark passages or places, evenness of temperature, the home as the frame of its inmates, for rich and poor alike will appreciate these qualities."

Venturi Robert & Scott Brown Denise

Following architectural studies at Princeton University, New Jersey, Robert Venturi (born 1925) held a scholarship at the American Academy in Rome from 1954 to 1956. He then went on to work in the architectural offices of Eero Saarinen and Louis I. Kahn, until 1958, as well as lecturing at Princeton and Yale Universities. Of particular note among his early works are the "Guild House" home for the elderly in Philadelphia, which he built with fellow architects Cope and Lippincott (1960–1963), and the house for his mother in Chestnut Hill, Philadelphia (1962–1964). Both buildings became icons of post-modern architecture. From 1964 to 1987 he worked in partnership with John Rauch, and they were later joined by Denise Scott Brown, Steven Izenour and David Vaugham. Many different projects were completed, such as the Humanities Building of the State University of New York in Purchase (1968–1973), Dixwell fire station in New Haven, Connecticut (1967–1974), the Brant house in Greenwich, Connecticut (1970–1973), Franklin Court in Philadelphia (1972–1976), the extension to the Allen Memorial Art Museum in Oberlin, Ohio (1973–1976), the Institute of Scientific Information in Philadelphia (1970–1978), with its façade of mosaic-like colored wall panels, and the Gordon Wu Hall of Princeton University (1980–1983). In contrast to the highly symbolic façades of his urban projects, which comply with his designation of "shelters with decoration", the holiday homes he built on Nantucket Island, Block Island or in Vail, Colorado have timber and shingle surfaces that fit in with the landscape, with windows cut out in unusual configurations that create surprising settings. In the publications "Complexity and Contradiction in Architecture" (1966), "Learning from Las Vegas", with Denise Scott Brown and Steven Izenour (1972), and "A View from the Campidoglio" with Denise Scott Brown (1985), Venturi reinforced his view that even modern building work exists through historical references and has to acknowledge the messy everyday architecture of "Las Vegas-style" strips as a competing reality.

GUCCI

GUCCI

CELINE

reflect.

the most emotionally intense color + it stimulates
heartbeat and breathing + increases blood circulation
self confidence + decreases lethargy and
helps ease chronic pain + may lower the frequen
causing patients to be less excited a
reduce pain + stimulates the automatic nervous a
circulatory systems + stimulates the liver + reduces infla
and swelling + supports the building of the blood

JALAPEÑO
Google

GUAJILLO
JALAPEÑO

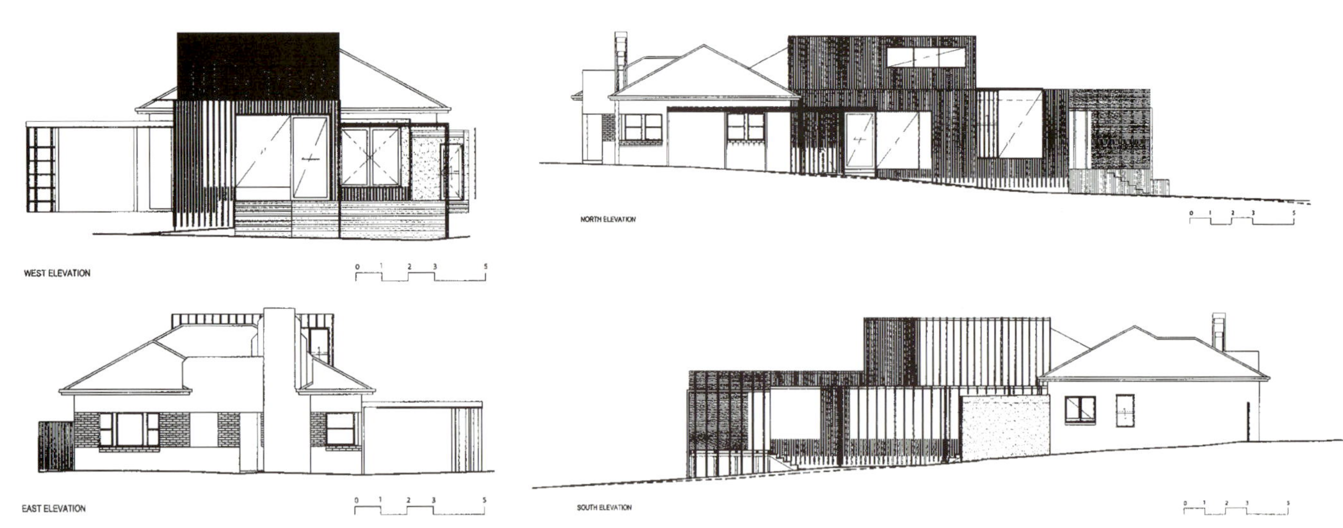

WEST ELEVATION
NORTH ELEVATION
EAST ELEVATION
SOUTH ELEVATION
0 1 2 3 5

House M

The project M, a residential building in the centre of Meran, is embedded in the quit area of Obermais. The concept of the design was it to play with transparent and solid surfaces what fallows fascinating insights and outlooks. The interior melts together with the outside space. The terrain flows through the building and finds his renewal in the pool- and meadow area. Because of a refined external design and the arrangement of the pool, lawn, garden and house the whole concept seems like a unity with seamless transition. The ground floor follows the slightly sloping ground like a staircase to get a large garden area. Because of engineering technology considerations the building is conceived as a compact volume with one underground and two upper floors. From the construction point of view the house is built as a concrete construction, furnished with upgraded insulation. The glass-facade, door and window elements are executed as 3 layered glazier.

Client
Privat
Location
Meran (BZ)
Program
Living, Villa
Living space
360 m²
Volume
1.390 m³
KlimaHouse
A
Photographer
M&H Photostudio
Credits
**monovolume architecture + design
(Patrik Pedò / Juri Pobitzer / Konrad Rieper)**
Coworkers
Luca di Censo, Sergio Aguado, Benjamin Gänsbacher

 in a modern way. The new extension is completely embellished with timber battens internally and externally, blurring the boundaries between the functional and the decorative as they extend beyond the façade and become shading devices, fences, privacy screens and pergolas.

The use of timber also references its Bayside location, offering a dialogue between the suburban frontage and a more coastal design approach to the rear, referencing its dual context.

Designer
Fiona Dunin
Company
fmd architects
Project Location
Melbourne, Australia
Project Team
Fiona Dunin, Kathrine Peasley, Rob Kolak, Alex Peck
Photographer
Peter Bennetts

Grenadier House

With the backyard now a usable green space, this house accommodates a double carport and garage below the main living quarters of the house at the front of the lot, eliminating the need for a lengthy driveway. The garage creates a split-level concept, with four stories at the back of the house and three at the front. The main living quarters are slightly elevated above street level, allowing sun into the south-facing facade while maintaining privacy within the dwelling. At the rear, the main living quarters extends into the backyard with large window and door openings. The top floor master suite overlooks the rooftop patio half a story below while maximizing on desired solar gain.

Design Agency
Altius Architecture, Inc.
Designer(s)
Graham Smith & Logan Amos

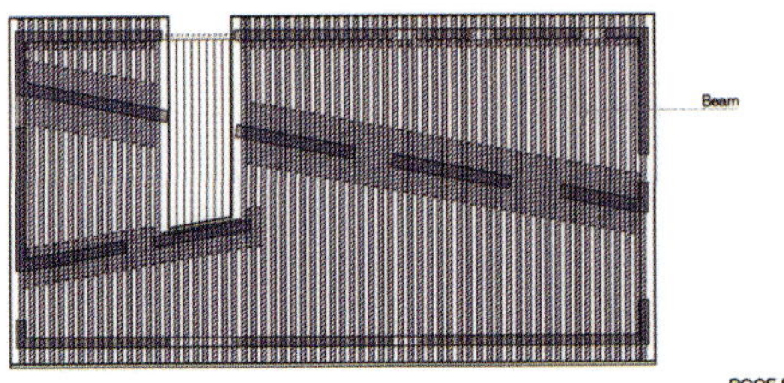

Beam
ROOF PLAN

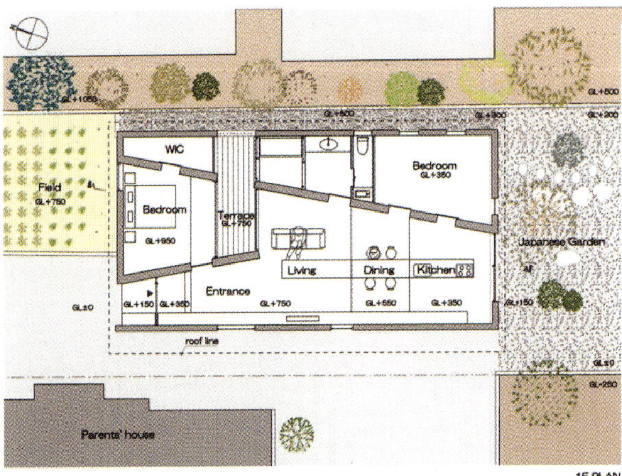

Field
GL+750
WIC
Bedroom
GL+950
Terrace
GL+750
Bedroom
GL+350
Japanese Garden
Living
Dining
Kitchen
Entrance
GL±0
GL+150
GL+350
GL+750
GL+550
GL+350
GL+150
GL+1050
GL+500
GL+500
GL+300
GL+200
GL±0
GL-250
roof line
Parents' house
1F PLAN

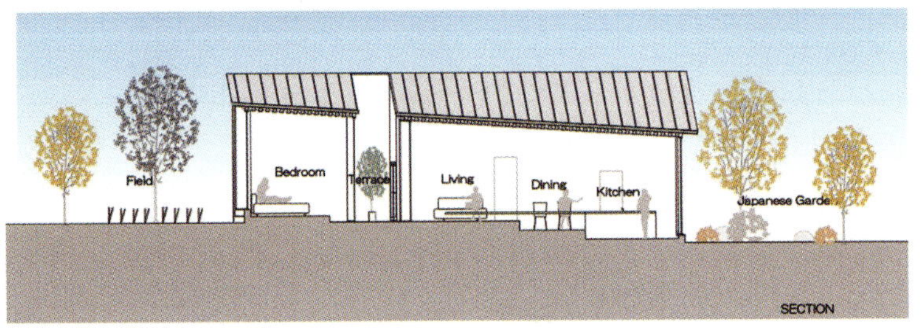

Field
Bedroom
Terrace
Living
Dining
Kitchen
Japanese Garden
SECTION

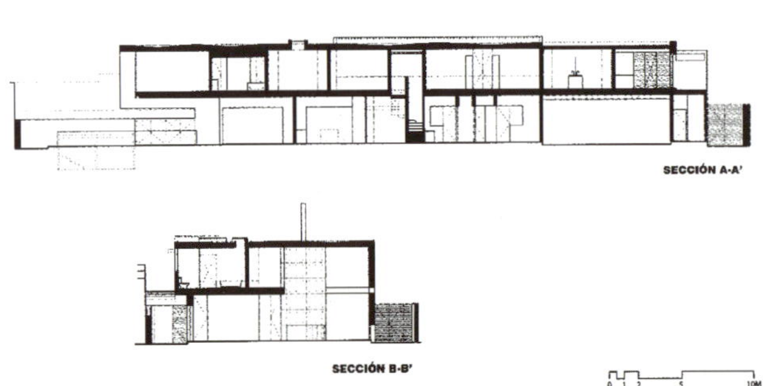

SECCIÓN A-A'
SECCIÓN B-B'
0 1 2   5   10M

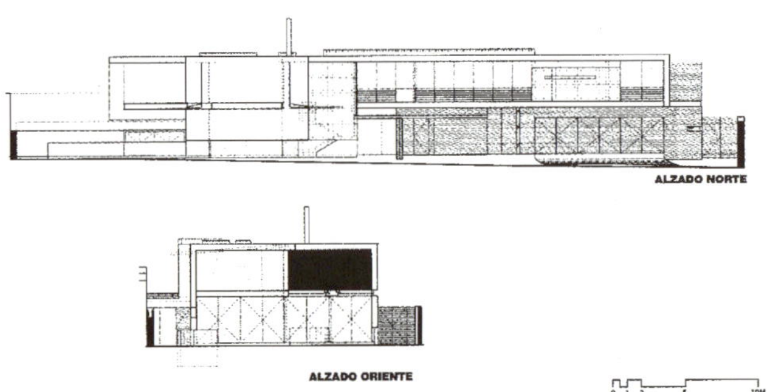

ALZADO NORTE
ALZADO ORIENTE
0 1 2   5   10M

MP apartment

The MP apartment was originally built in order to enjoy the landscape of Caldonazzo lake. The apartment, surrounded by gardens and orchards, has a beautiful view on the lake.

The furniture project is based on the observation of the surrounding landscape and the analysis of the actions that take place there. The furniture evokes the need to find a new balance between opposites, to regenerate body and mind. Natural vegetal and mineral materials are used, such as oak, declined as a coating for floors and walls, and lime plaster, which can be both practical and comfortable. Vegetable and minerals materials are to be found in the garden, which has become the place where the wild and the domestic landscape melt into one another.

In the living area, the planning define a unique space: as if it were a totem, a central piece of furniture, which includes the fireplace, has become a functional pin, but it is also useful and versatile, almost like a boat locker, like those that, on sunny days, appear on the lake.

Company
Burnazzi Feltrin Architects
Creative Director
Elisa Burnazzi and Davide Feltrin
Designer
Elisa Burnazzi, Davide Feltrin

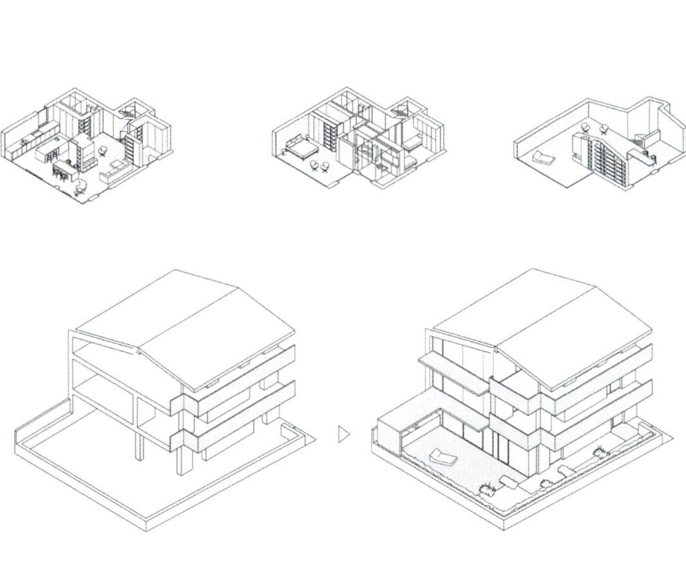

Pojagi House

Traditional Korean patchwork panels called pojagi serve here as sliding doors in a modern take on the classic Japanese design concept of "tsuzukima," a large tatami-mat-lined space that can be divided as needed with screens. The moveable pojagi screens carve out a diversity of interior scenes depending on their layout, adding flexibility to the space. Together with 150mm-square wood colonnade walls, they create a warm, relaxed atmosphere within the simple layout.

Design Agency
MDS Co.Ltd
Photographer
Toshiyuki Yano
Designer(s)
Kiyotoshi Mori/Natsuko Kawamura

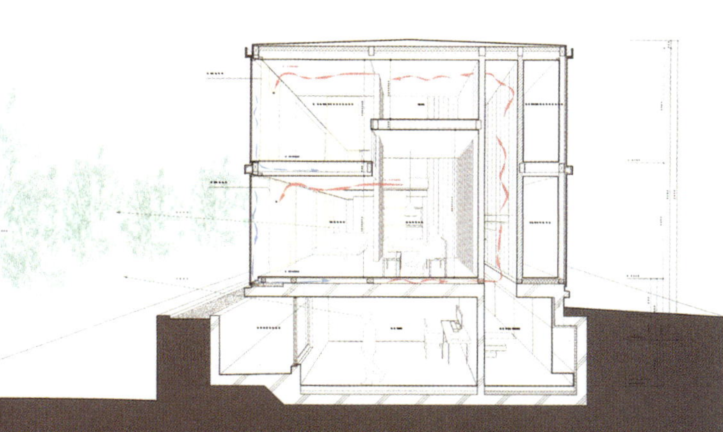

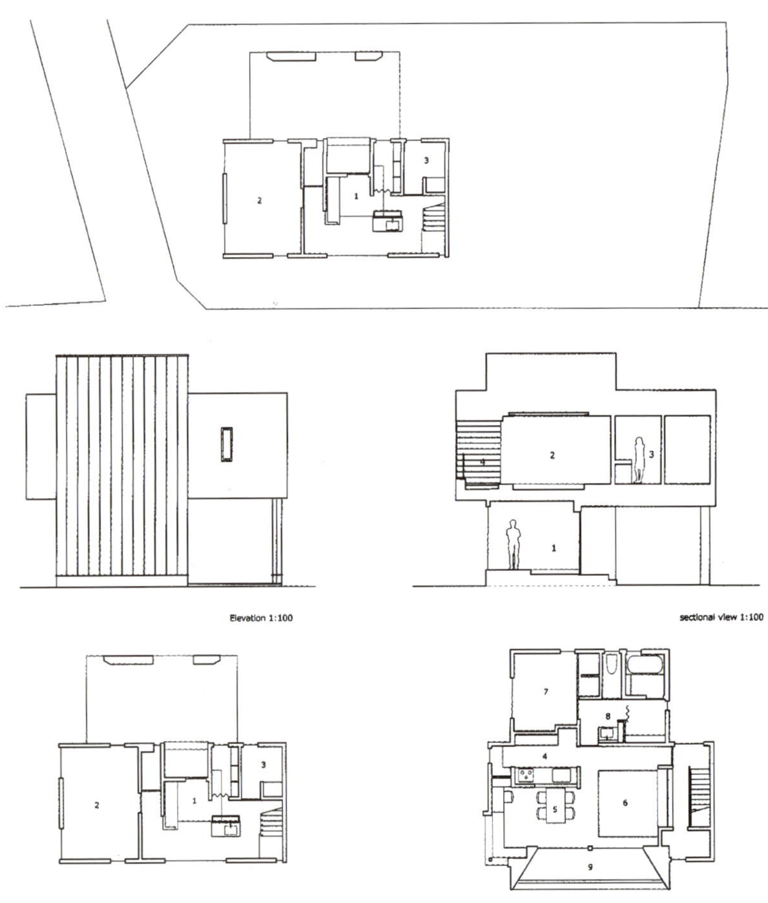

Elevation 1:100

sectional view 1:100

1st floor

2nd floor

Vigo University Campus 1999 / Vigo, Spain
photo: ©Duccio Malagamba
Igualada Cemetery 1995 / Barcelona
Olympic Archery Pavilions 1991 / Barcelona
photo: ©Duccio Malagamba

Freedom Tower / New York

Sears Tower   1974 / Chicago, Illinois
photo: ©Timothy Hursley

John Hancock Center   1970 / Chicago, Illinois
photo: ©Timothy Hursley

Jin Mao Tower   1999 / Shanghai
photo: ©Hedrich-Blessing

Burj Dobai Tower / Dobai, United Arab Emirates

CATV

Tezuka Architects

Below: "House to Catch the Forest", Chino, Japan, 2004

Opposite page: "Wall-less House", Tokyo, Japan, 2001

Takaharu Tezuka (born in 1964) and Yui Tezuka (born in 1969) both graduated in architectural studies from the Musashi Institute of Technology in Tokyo. Takaharu Tezuka gained a Masters degree at the University of Pennsylvania before he found a position with Richard Rogers. Yui Tezuka continued her studies at the Bartlett School of Architecture in London. In 1994, they co-founded the architectural practice Tezuka Architects in Tokyo. Both are in the teaching profession. With the "Wall-less House" in Setagayaku, Tokyo (2001), the whole edifice is supported by a stable core and extremely thin pillars. The ground floor manages without any walls at all and, with the glass walls pushed aside, merges directly into the large garden. On a small plot of land in this densely populated neighbourhood, the "Thin Wall Office" in Shibuyaku, Tokyo (2004) makes efficient use of the available space with thin walls and the permissible roof height. Nine-millimeter-thick sheets of steel and narrow pillars support the structure. The glass elevations to the front and rear give an insight into office life. The basement, however, together with the bathroom and kitchen, remains out of sight. The Engawa House in Adachiku, Tokyo (2003) was erected as an additional building in the garden of a detached family house. The façade extending to the long, narrow courtyard and the main house is completely glazed, so that the building can be transformed into a 16-meter-long, roofed veranda, with the help of the nine sliding glass doors. Like a display window, all the rooms are clearly visible from the courtyard – even the bathroom and bedrooms. A narrow band of ribbon windows, which extends over the entire length of the rear of the building, guarantees enough daylight. The "House to Catch the Forest" in Chino in the Nagano prefecture (2004) is mounted on three plinths in a pine forest. The acutely raised, glazed frontage of this holiday house offers beautiful views of the surrounding treetops. The steeply sloping roof prevents too much snow accumulating.

Utzon Jørn

Below left: Bagsvaerd church, near Copenhagen, Denmark, 1969–1976

Below right and opposite page: Sydney Opera House, Australia, 1957–1973

Jørn Utzon (1918–2008) was a graduate of the Academy of Art in Copenhagen. Until he opened his own office in Copenhagen in 1950, he worked with Paul Hedqvist in Stockholm and Alvar Aalto in Helsinki. He then travelled, gathering ideas, throughout Europe, North Africa and America, where he visited, among others, Frank Lloyd Wright. In the 1950s, Utzon built his own house in Hellebæk, Denmark (1952) and the Kingohusene project in Helsingør, Denmark (1956–1960), a serpentine formation of housing blocks. These flat, single-family units constructed of yellow brick are closed on the street elevation, opening out onto enclosed garden courtyards. Among his other works are the terraced houses in Fredensborg, Denmark (1962–1963) and the Birkehoj houses in Elsinore, Denmark (1963). The Sydney Opera House (1957–1973), located on a peninsula in the harbour, consists of striking concrete domes shaped like mussel shells that are completely clad in bright tiles: these shells are staggered across two platforms, under which the main halls are accommodated. The completed interior of the opera house differed to some extent from the designs of Utzon, who only supervised the construction for a few years. He also designed Bagsvaerd church in Copenhagen (1969–1976): a stepped structure with glazed triangular louvers that features undulating ceiling panels within. The Utzon house in Porto Petro, Majorca (1971–1972) is once again a collection of small individual buildings. Utzon was also frequently involved in the design of furniture and lighting.

Steiner Rudolf

Rudolf Steiner (1861–1925) began his studies at the technological university in Vienna in 1879. From 1883 onwards he was intensively occupied with the study of Goethe's scientific writings. For an understanding of his aesthetics, the paper "Goethe as father of a new aesthetic" is of central importance: the purpose of art is to bridge the gap between the physical and the metaphysical. In 1902, Steiner joined the theosophy society and started busying himself with the concrete translation of his ideas. In 1907, he designed column capitals, which were later used in the first Goetheanum. By his own account, he produced a model of a Rosicrucian temple in Malsch in 1908 and 1909. Between 1911 and 1913, he was in charge of the planning of the "Johannesbaus" in Munich, the completion of which was defeated because of the resistance of the authorities and churches. In 1913, Steiner founded the Anthroposophical Society. The Goetheanum buildings,

which were completed in Dornach over the following years, were to offer the Anthroposophists rooms for their work. Resuming his interest in Goethe's ideas about the metamorphosis of plants, in his first Goetheanum (1913–1920) Steiner tried to translate that "inner growth pattern of nature" described by Goethe into architecture. He produced drawings and models, which were used at lectures on architecture. Over a concrete base, a wooden structure with two domes rose up, which was primarily meant to suggest a place of safety. As in the mother's womb, which represented the 'elemental house' for Steiner, there were no straight angles, as everything was rounded inside the building. The roof, which was made of Norwegian slates from Voss, curved protectively over the building. For Anthroposophists who settled in Dornach, Steiner designed residential buildings such as the Duldeck house (1915–1916). Stylistically, it occupies a mid posi-

tion between the first Goetheanum and the second. It seems a little over-theatrical. Primarily it gives this impression on account of the massive roof, which is reminiscent of Gaudí's apartment blocks in Barcelona. The first Goetheanum was still strongly influenced by Jugendstil (Art Nouveau) and so the second, which Steiner designed in 1923 after the first Goetheanum had gone up in flames, was a masterpiece of Expressionism. Steiner planned the second Goetheanum in concrete. He made reference to it – the external appearance should generally result from moulding the material – and promoted a new style of concrete cast into forms. As Steiner died before its completion (1924–1928), the second Goetheanum was built in accordance with one of his finished exterior models. The monumental nature of the building is one of the characteristics of the remarkable architectural style. At the same time, the whole building is cast throughout into such intricate forms due to the curvature of the exposed concrete surfaces that, with the changing light, new shapes constantly appear. Steiner was regarded as the founder of Anthroposophist architecture. It has spread to many parts of the world in different forms, yet is always committed to Steiner's basic ideas, which is to say, the abandonment of straight angles and in their place the use of curves, polygons and sloping planes as well as the emphasis on detailed craftsmanship. In essence, Anthroposophist architecture is similar to organic building.

"Until now, the artistic concept of building has been that of the idle, lifeless, mechanical aspect; but now the artistic concept of building is starting to become the concept of 'eloquence', the concept of inner liveliness, and such a concept carries us away with it."

Scarpa Carlo

Carlo Scarpa (1906–1978) studied at the Accademia di Belle Arti in Venice until 1926, whilst also working in the practice of Vincenzo Rinaldo. He then took up a position as assistant to Guido Cirilli at the newly established Istituto Universitario di Architettura in Venice. One of his first projects was to design the Florence branch of the Maestri Vetrai Muranesi Cappellin & Co. glass-manufacturing firm. Over the course of the next 30 years, Scarpa remained a frequent designer for the Venice Biennale. Of particular note in this respect was his design for the 1948 Paul Klee exhibition and pavilion for the 1950 Art Book exhibition. Between 1955 and 1961, he worked on the Veritti residence in Udine. He was particularly praised for his work on the conversion of the Museo Correr (1953–1960) and the Ca'Foscari (1954–1956), both in Venice. At the same time, he was collaborating with Ignazio Gardella and Giovanni Michelucci in designing the first six rooms of the Uffizi in Florence (1954–1956) and planning the extension to the Canova Museum in Possagno (1955–1957). By introducing unorthodox window solutions, openings and light shafts,

Scarpa found clever ways of introducing light into the gallery to display the sculptures to their best possible advantage. One of his subsequent projects was to design the Olivetti display and sales rooms in his hometown of Venice (1957–1958). In 1962, he was appointed Associate Professor of Interior Architecture. Two years later, he completed the rebuilding of the Museo Castelvecchio in Verona (1956–1964): a project in which he not only marked the difference between his own conversion work and the original, but also the previous rebuilding phase by "peeling back" the roof and exterior layers. After spending time travelling in the USA and Japan, he was commissioned by Rino Brion to design a family tomb complex on the periphery of San Vito d'Altivole cemetery near Treviso (1969–1978). On a plot comprising an area of 2000 square meters, Scarpa designed an entire artistic masterpiece comprising gateway, chapel, tomb and medita-

tion pavilion, enclosed by a sloping wall reminiscent of a bastion. In 1972, he became head of the Istituto Universitario di Architettura in Venice, where he lectured until 1977. In 1973 he started his last major project, the Banca Popolare in Verona, in collaboration with Arrigo Rudi, who completed the project in 1981. Scarpa died without ever quite managing to gain full recognition as an architect.

"Frank Lloyd Wright's work swept me away like a wave."

"Always design a thing by considering
it in its immediate, larger context –
a chair in a room, a room in a house,
a house in an environment, an
environment in a city plan."

Safdie Moshe

Israeli-born architect Moshe Safdie (born in 1938) trained under Louis Isidore Kahn. Since 1964, he has lived in Canada, where in 1967 he designed his famous multi-family complex "Habitat 67" for the Montreal Expo: 158 rectangular, pre-fabricated boxes stacked on top of each other in a confused fashion, rising up to create the impression of a hill town. The flat roofs serve a dual purpose, as each one provides a balcony for a neighbouring flat. Different sizes of window, some of which are sited across the corners of the modules, lend the building an even more distinctive appearance. Safdie also realized the Yeshivat Porat Yosef Rabbinical College (1971–1979) as well as the Mamilla project (1972) in Jerusalem. Both building projects reflect the type of building design that is traditional for the region. Salt Lake City Library, Utah (1999–2003) was generously designed to serve other public functions. A public plaza, for example, was incorporated behind the five-story building on its triangular plot. One of its three façades has a curved wall that embraces the outdoor area, inviting visitors into the park. Safdie has also published several works, including "Beyond Habitat" (1970); "For Everyone a Garden (1974); and "Form and Purpose" (1982).

"Only originality born in the resolution of truly architectural issues contributes to great design."

SITE Projects Inc.

James Wines (born in 1932) studied at Syracuse University in Chicago and worked as a sculptor between 1955 and 1968 before founding, with Alison Sky (born in 1946), the multi-discipline organization SITE (Sculpture in the Environment) in 1970 in New York. Michelle Stone and Emilio Sousa also joined them. The goal of this collaboration was to combine art and architecture into an entity for which Wines coined the term "De-architecture". He was appointed a professor at the New Jersey School of Architecture in New York in 1975. For the Best Products business enterprise the SITE group redesigned several supermarkets, including the "Peeling Projects" building in Richmond, Virginia, (1972), where the brick façade peels away from the foundations. The "Indeterminate Façade" of the Houston showroom (1974) looks like a crumbling ruin. For the Notch Showroom in the Arden Fair Shopping Center in Sacramento, California (1977), the space for the entrance is made by the setting of the corner of the building on an extended set of rails. The project "Highrise of Homes" (1981) is intended to accommodate apartment blocks, including the garden area, in a high-rise skeleton, in order to save space. On the occasion of the World Exhibition in 1986 in Vancouver, SITE designed a 217-meter-long, undulating street called "Highway '86", which plunges into the sea at one end and towers up

Rogers Richard

Richard Rogers (born in 1933) studied at the Architectural Association in London and at Yale University, New Haven, Connecticut under Serge Chermayeff. In 1963, Rogers, his then wife Su and the Fosters founded Team 4, whose most important industrial construction was the Reliance control factory in Swindon, England (1967). That same year, Rogers represented British architects at the Paris Biennale for the second time. He also lectured at Cambridge, the Architectural Association and the London Polytechnic. In the late 1960s, he and his wife designed a lightweight, flexible building made of plastic rings; in 1971, he developed it further under the title of "Zip-up". He also built his own house in Wimbledon (1968–1969): a lacquered steel frame with plastic in-fills. In addition to teaching at Yale University, the Massachusetts Institute of Technology and Princeton University, from 1969 he collaborated with Renzo Piano on a number of projects that were never realized. In 1971, they won the competition for the Centre National d'Arts et de Culture Georges Pompidou in Paris (1971–1977). The appearance of the six-story complex is determined by the exposed technology, including construction grids, brightly colored supply elements and transparent traffic tubes. After separating from Piano in 1977, Rogers moved his office back to London, where he designed and built the Lloyd's skyscraper (1979–1986). At the same time, he also created the Inmos microprocessor factory in Newport, Gwent, South Wales; the PA Technology Center in Princeton, New Jersey (1982–1985); and the PA Technology Centre in Cambridge, UK (1975–1983). Other projects include the European Court of Human Rights in Strasbourg (1989–1995) and the courthouse in Bordeaux (1993–1996). Closed, cedar-clad spheres rise up from behind the long, glass façade, penetrating the undulating roof and enclosing the courtrooms. At one point, the roof of the Welsh Assembly in Cardiff (1998–2005) reaches down into the assembly hall and illuminates it in a grandiose gesture.

"We have reached the future, but we are only just beginning to see its influence on architecture."

"Architecture needs to give people shelter, to exalt their life on earth and to fulfil their belief in the nobility of their existence."

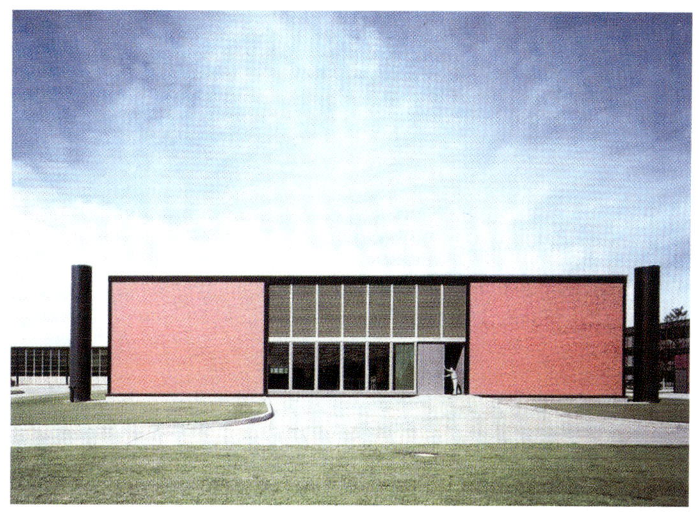

Opposite page: War Memorial Center, Milwaukee, Wisconsin, 1953–1957

Above: General Motors Technical Center, Warren, Michigan, 1948–1956

Below: Jefferson National Expansion Memorial, St. Louis, Missouri, 1947–1968

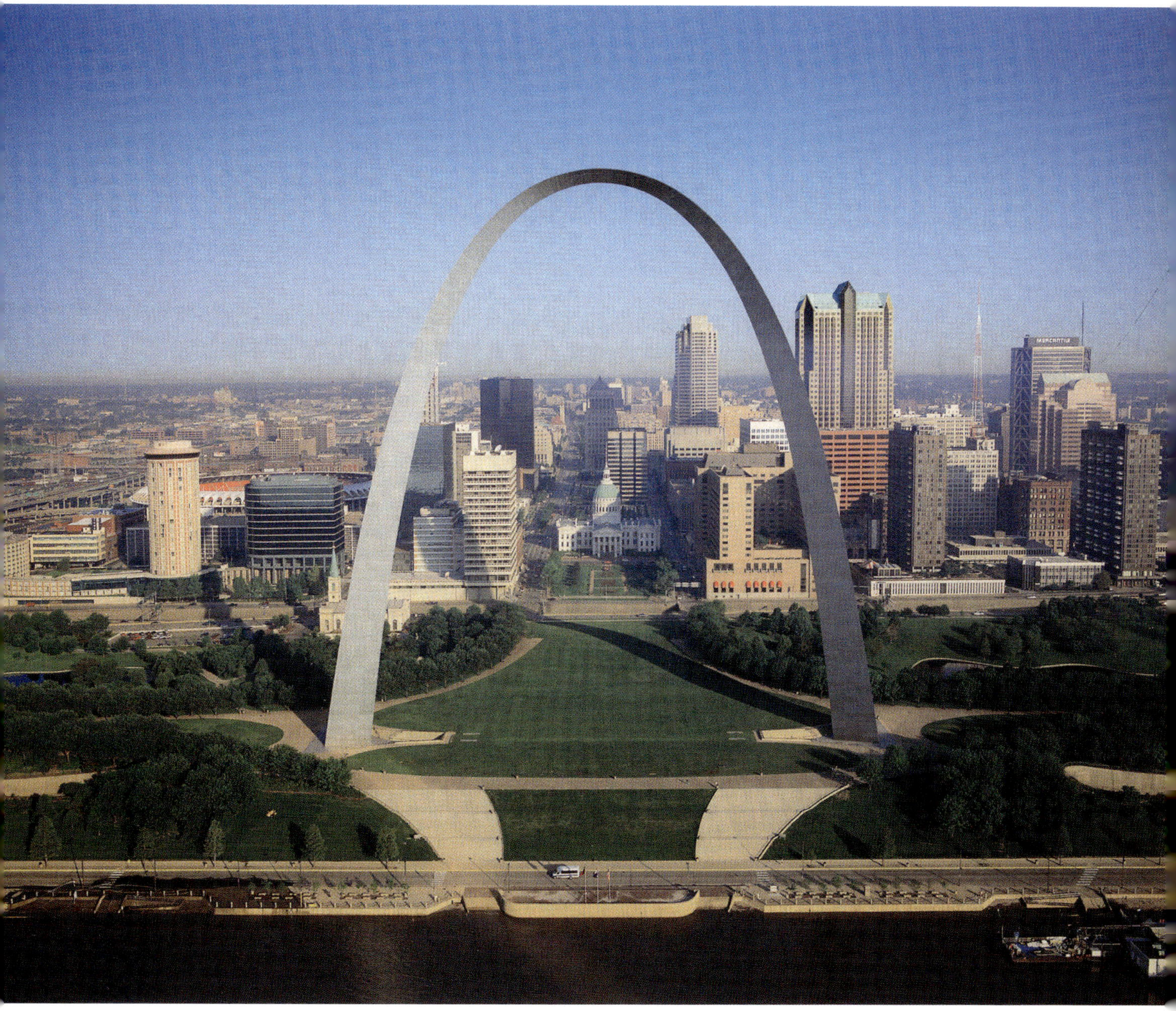

Predock Antoine

Polytechnic University in Pomona (1987–1992). The stone tower, triangular in design, which houses the Social Sciences and Humanities faculties, forms a distinctive landmark on the horizon. Some of his other projects include the Euro Disney Santa Fe Hotel in Marne-la-Vallée, France (1992), featuring tiered, pastel-colored façades, as well as the Ventana Vista Elementary School in Tucson, Arizona (1992–1994). The latter, constructed in natural stone and echoing the hues of the desert and nearby mountains, almost melts into the surrounding landscape. In addition to numerous individual residential houses, he also produced a large number of public buildings. In many cases, his structures became landmarks in their own right: for example, the Spencer Theater for Performing Arts in Ruidoso, New Mexico (1994–1996) and the Green Valley Center for Performing Arts in Arizona (1996–2004). The Gateway Center of the University of Minnesota in Minneapolis (1996–2004) was designed and built in collaboration with Korsunsky Krank Erickson Architects Inc. In form and color, it suggests the different geological layers of a cliff-face. Austin City Hall, Texas, was realized between 1999 and 2005.

"The enigmatic quality of the desert is crucial to the spirit of my work... The lessons I've learned here about responding to a place can be implemented anywhere."

Antoine Predock (born in 1936) attended the University of New Mexico in Albuquerque and Columbia University, New York. Since 1967, he has been the principal of his own firm in Albuquerque as well as lecturing at the University of California at Los Angeles and at California State Polytechnic University in Pomona. A distinctive feature of his architectural style is his preference for unusual stereometric forms and for integrating local traditions. The natural colors, pyramid and cube shapes, which characterize the structural elements of the Fuller house in Scottsdale, Arizona (1984–1987) are sensitively blended into the desert landscape. Water flows right through the house and into a circular pool. The Nelson Fine Arts Center of Arizona State University at Tempe (1985–1989) with its smooth brick and concrete façades was likewise built during the eighties, as was the cone-shaped, copper-clad American Heritage Center in Laramie, Wyoming (1986–1993). Working in collaboration with Gensler & Associates, Predock also produced a multi-functional building for California State

Opposite page: Von Sternberg
house, San Fernando Valley,
California, 1935–1936

Right: E. J. Kaufmann house,
Palm Springs, California,
1946–1947

Melnikov Konstantin

Konstantin Stepanovich Melnikov (1890–1974) first trained as a painter of icons. He then studied at the School of Painting, Sculpture and Architecture in Moscow, receiving his diploma in painting in 1914 and architecture in 1917, when he also designed an automobile factory in Moscow. After the revolution, he worked in the architectural workshop of the Moscow Soviet in the USSR and joined the department of architecture of the People's Commissariat for Education. From 1920 to 1929 he was professor at the VKhUTEMAS (higher artistic-technical workshops). In 1923, Melnikov developed plans for a workers' housing area; in 1924 he designed the market at Sucharevsk, and in 1924 an exhibition pavilion for the Machorka tobacco company, also in Sucharevsk: its rectangular wooden frame was an important step in his research into the possibilities of design. Melnikov built the Soviet pavilion for the International Exhibition of Decorative Arts and Modern Manufactures in 1925 in Paris: Alexander Rodchenko supplied the color concept for the wooden con-

struction. A diagonal staircase was cut through the cube up to the top floor, while the ceilings and external walls were made almost entirely of glass. His design of the same year for a multi-story car park over the Seine in Paris was not realized. His own home in Moscow (1927–1929) consisted of two, intersecting, cylindrical towers: one with a large glass section, the other with hexagonal windows. Work on six workers' clubhouses commenced at the same time: the club for the Svoboda factory ("Liberty", 1927–1929), the "Frunze" club (1927), the Rusakov club (1927–1929), Pravda club (1927–1928), the "Kauchuk factory" club (1927–1931) and the Burevestnik club ("Stormbird", 1928–1930). Melnikov built four garages in Moscow between 1926 and 1936. In 1929, he designed a memorial for Christopher Columbus in Santo Domingo, which was never realized, as well as a union theater and the People's Commissariat for Moscow Heavy Industry in 1934. His independent feeling for form led to a criticism of formalism, and he subsequently re-

ceived no commissions apart from a few interior conversions; from 1937 to 1953, he was not even allowed to use the title of architect.

"I was given the title of architect, and I entered a profession that stood on the edge of an abyss. Why does my work create such a strong curiosity, blended with concern? What gives rise to the distaste and fear of the extraordinariness of this work, and why does one experience a feeling of freshness when one is more familiar with it? I know: this century, I was called to breathe new life into the dying sense and speak in a new language of architecture."

SUN ROOF ▶

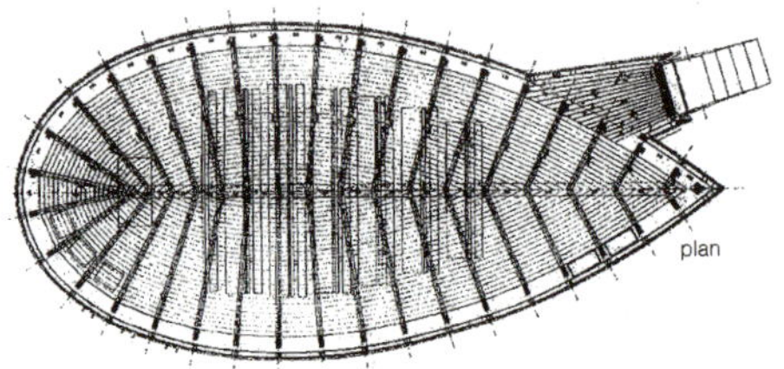

Deep in the mountains of eastern Switzerland lies the small village of Vals, where hot springs have been bubbling up for over 100 years. After a hotel was built, and visitors came to test the waters, the town really came into its own. But what has made it even more famous is a bathing facility called Thermal Vals that was designed by Peter Zumthor.

In the singular Swiss landscape, which unfolds against a backdrop of precipitous mountain peaks, Thermal Vals stands submerged on the lot's western slope. The mountain greenery slants downward to become one with the roof of the building. The interior and exterior walls of the structure, about half of which is underground, are made of thinly sliced layers of gneiss from the Vals area. The visibly dense, slate-blue texture is enough to send a shiver through any architect. Only Zumthor could create such a charming appearance simply through his handling of a material. I also understand the attraction of the building, having visited and stayed in the hotel three times and savored the architecture, springs, bar, Zumthor's room, and the landscape. According to Zumthor's wife, who serves as the director of the facility, "In terms of healing power, three days here is equal to three weeks at an Italian resort." My personal feeling, though, is that nothing can compare with the cool, silent inner and outer walls, and the magnificent sight of the remaining patches of snow framed by the terrace.

Peter Zumthor was trained as a furniture craftsman by his father, a professional artisan. After attending art school in Basel, Zumthor graduated from New York's Pratt Institute with a degree in architecture. Needless to say, this explains why Zumthor has such a wealth of knowledge about materials and such a subtle sense of expression. The products of these special talents are Thermal Vals and his next masterwork, Saint Benedict Chapel.

The small church, a wooden structure with a leaf-like planar form, is surrounded by a shingle-thatch exterior wall. Zumthor has been quoted as saying, "By receiving the site of the building, the material becomes a poetic existence." The color of the simple, shingle wall has changed by being exposed to the elements, giving it an appearance that is suited to a rustic, rural chapel. Both the hotel and the chapel are fine examples of Zumthor's poetic way with stone and wood – natural materials which he is especially adept at using –, but for his next important work, the Art Museum Bregenz, the architect proved that he was just as skillful at dealing with artificial materials such as glass and concrete.

Residential Building at Casares & Gelly   2002 / Buenos Aires, Argentina
Bard College, Center for Science & Computation / Annandale-on-Hudson, New York
Leicester Theater and Performing Art Center / Leicester City, UK
Tampa Museum of Art / Competition

Rafael Vinoly is one of the most celebrated architects of the era. Although completely unknown in Japan at the time, he managed to win the Tokyo International Forum competition, the first International Union of Architects-approved project in Japan. And not only did the finished building, a huge structure, become one of Vinoly's representative designs, it drew many to his rich body of work.

Born in Uruguay in 1944, by the time Vinoly was 20, he had already become a founding partner in one of the largest architectural firms in Latin America, Estudio de Arquitectura. While working at the company's headquarters in Argentina, he decided, at the age of 34, to relocate to New York. After serving as a guest lecturer for a brief period at the Harvard University Graduate School of Design, he settled in New York in 1979, and established his own firm there in 1983.

In addition to the Tokyo International Forum, I have had the opportunity to visit Vinoly's Kimmel Center for the Performing Arts, David L. Lawrence Convention Center, Lehman College Physical Education Facility, Princeton University Stadium, and the Lewis-Sigler Institute for Integrative Genomics at Princeton University. All of these buildings are large-scale works and all of them feature a unique form and structure.

Judging from what I have seen, the charm of Vinoly's architecture lies in his dynamic structures. With a length of 208 meters and a huge roof topping the 57.5-meter building, the Tokyo International Forum is a particularly impressive example. The breathtaking roof, supported by a series of ship-bottom-shaped, cast-iron ribs, is part of a tremendous structural system bolstered by two huge spindles with a maximum diameter of 4.5 meters.

The Kimmel Center, located in Philadelphia, also boasts an absolutely stunning structural system. The large space within the gigantic, glass barrel vault that covers the site, which occupies one entire block, is created by a cornice-shaped Vierendeel truss. Within it are two concert halls in separate buildings. The urban environment has been taken inside the transparent vaulted building to create a "city within the architecture."

The Lawrence Convention Center in Pittsburgh is equipped with a suspension-bridge-type roof that was inspired by the city's many bridges. The roof element, with a long streaming appearance similar to a bolt of cloth, gives the riverside landmark its unique appearance.

David L. Lawrence Convention Center   2003 / Pittsburgh, Pennsylvania
photo: ©Roman Vinoly

Jazz at Lincoln Center   2004 / New York
photo: ©Brad Feinknopf

Penn State University, School of Information Sciences & Technology
2003 / University Park, Pennsylvania
photo: ©Joseph David

University of Chicago, Graduate School of Business   2004 / Chicago, Illinois
photo: ©Brad Feinknopf

Carl Icahn Lab of the Lewis - Sigler Institute for Integrative Genomics at Princet
2003 / Princeton, New Jersey
photo: ©Synectics

Nasher Museum of Art, Duke University
2004 / Durham, North Carolina
photo: ©Brad Feinknopf

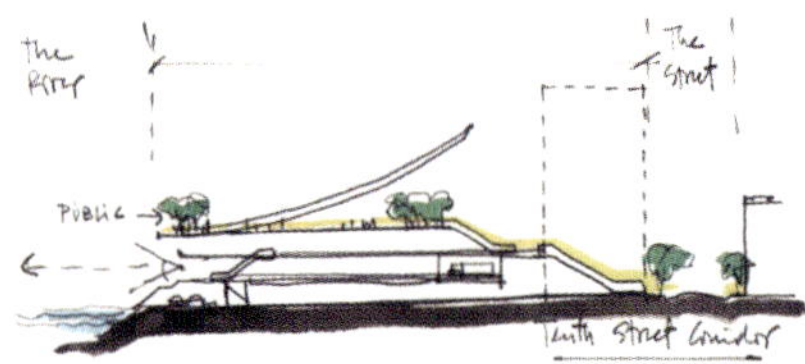

To the immediate left of the main entrance to New York's Museum of Modern Art (MoMA) stands the American Folk Art Museum. With a facade that consists primarily of a metallic material, the building creates a fantastic first impression. Since the building was completed in 2001, its design, by Tod Williams Billie Tsien, has been hailed not only in the American architecture world, but in architecture magazines around the world. Compared to MoMA, the twelve-meter frontage of the building is small, but the sculpted exterior is more eye-catching than the flat facade of MoMA. The form is an abstraction of the "hand," the tool which creates folk art, and is covered with panels of tomabsil (a form of whitish bronze) used for propellors and gun barrels.

Even more popular than the building's facade is its elaborate interior. The key to the Williams-and-Tsien style is the selection of a material which is suitable to the quality of the space, and the technique used to distribute the material within it. Williams explains, "The special characteristic of our architecture lies not in the act of looking at a building, but rather in experiencing it. Therefore, the choice of materials and details as well as the construction method and the use of the building become extremely important."

In the same way, the office wing, laboratory, and auditorium at the Neurosciences Institute in La Jolla are fitted with a variety of marvelous materials and arranged in an organic way around a courtyard. As the lot is sloped, the approach leads to the upper section of the laboratory, and by looking over the unique frosted-glass balustrade, one is able to take in all of the building's features including the courtyard – an exquisite touch.

The auditorium, open to the neighboring residential area for concerts, is a particularly gorgeous place with Canadian redwood running through the interior, and an abundance of Texas limestone and Italian marble used in the walls and floors.

Born just outside of Detroit, Williams studied at Princeton and Cambridge before returning to Princeton for a graduate degree, but even as a child he dreamed of becoming an architect. Higher education merely served to finalize his goal. Tsien, meanwhile, is from New York. Initially, Tsien specialized in graphic design and though she liked painting and art, she had no interest or knowledge of architecture. But on a friend's recommendation, she entered the School of Architecture at UCLA and gradually

RAFAEL VINOLY

USA

Born in Uruguay in 1944. At the age of 20, had already become a founding partner in one of the largest architectural firms in Latin America. Relocated to the U.S. in 1978, and served as guest lecturer at the Harvard University Graduate School of Design. Settled in New York in 1979 and established own firm there in 1983. After opening a second office in London, the firm now employs over 170 people.

Inholland University   2000 / Rotterdam
Alphen aan den Rijn City Hall   2002 / Alphen aan den Rijn, The Netherlands
Crawford Municipal Art Gallery   2000 / Cork, Ireland
Royal Netherlands Embassy in Poland   2004 / Warsaw
Luxury Housing Mauritskade   2002 / Amsterdam

Main Building and Auditorium, University of Leipzig / Augustusplatz, Germany
Russian Avant-Garde Residential Complex / Moscow
Mahler 4 Office Tower / Amsterdam
Natural History Museum Rotterdam   1996 / Rotterdam
photo: ©Synectics
Kroyers Plads Housing Complex / Copenhagen

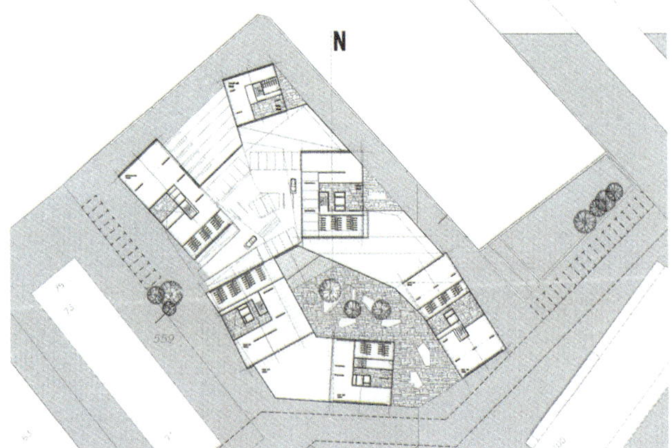

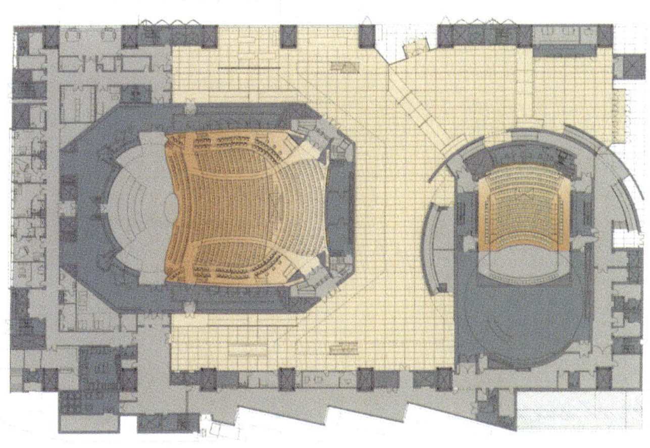

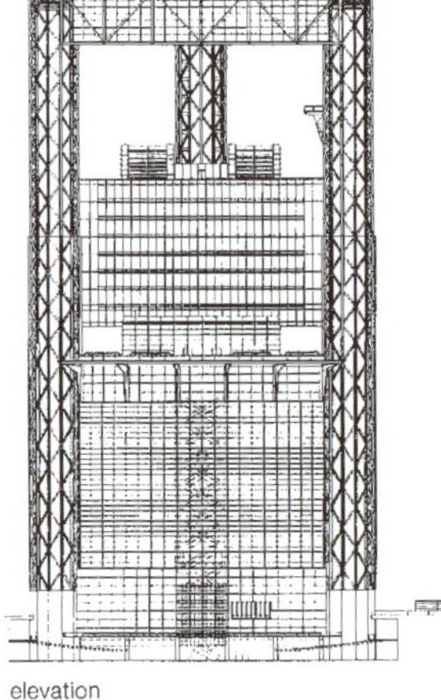

elevation

Samsung Jong-ro Tower   1999 / Seoul
photo: ©Kim Jung Oh

Tokyo International Forum   1996 / Tokyo

Lehman College Physical Education Facility
1994 / Bronx, New York
photo: ©Peter Margonelli

Princeton University Stadium
1998 / Princeton, New Jersey
photo: ©Michael Moran

Boston Convention & Exhibition Center
2004 / Boston, Massachusetts
photo: ©Brad Feinknopf

Brunete

Designer Brunete Fraccaroli
Location São Paulo – Jardim Paulista
Area 200m²

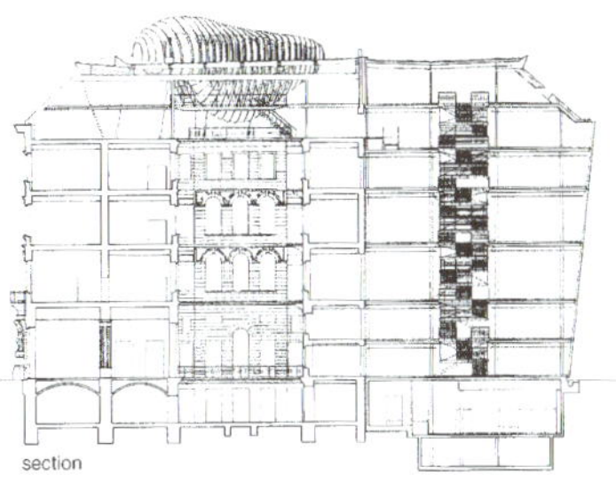

section

Erick van Egeraat started his career as one of the founding members of a young Dutch architecture group called Mecanoo. After leaving the group in 1995, Egeraat established his own firm in Rotterdam called Erick van Egeraat Associated Architects (EEA). As of 2005, after ten years of steady growth, the firm had offices in London, Prague, and Budapest, and had turned into a major practice employing a staff of some 100 people. What is especially notable about the firm is its success in Eastern Europe.

Capital City (or Gorod Stolits) is a huge project that includes office and residential spaces, and a retail area. The structure consists of two low rises, a cone-shaped dome, and two super high rises, one 49-stories-tall and the other 61-stories.

The low-rise section houses a shopping and entertainment area, with more retail outlets in the mid-rise section and a combination of offices and residences on the higher levels of the structure. The most surprising thing about the work is the cutting-edge architectural expression used for all three buildings. To make the two super high-rise buildings appear less massive, van Egeraat altered the position of the diagonal axis line to make it look slightly off-center. The upper part of the facades also have a number of overhanging sections and varying levels, and through van Egeraat's use of a combination of horizontal, vertical, and square openings, he succeeded in creating a powerful expression, which includes a strong element of movement. Movement of this type can often be found in van Egeraat's work.

In Budapest, van Egeraat designed the ING Bank and the NN Hungary offices, and the ING Head Office. He renovated and expanded the original 19th-century building, covered the courtyard with a glass roof, and created a boardroom which resembles a whale. The unique whalelike structure is especially interesting for its strong sense of movement. The facade of the building is segmented into countless vertical stripes, which look as if they're undulating. Because this undulating effect adds dynamism to the work, it is called a "moving facade." Similarly, the cantilevers at the edges of the raised Alphen aan den Rijn City Hall make it look like some kind of animal raising its head. Moreover, the comical appearance of Popstage Mezz in Breda, which according to one's perspective resembles a seal clothed in metalic sheets, is truly amusing. Despite the growth of his firm, van Egeraat still feels the need to be personally involved in the

Kurokawa Kisho

Nagakin Capsule Tower, Tokyo,
Japan, 1969–1972

Kisho Kurokawa (born in 1934) attended the universities of Kyoto and Tokyo and was employed by Kenzo Tange. His office has been in existence in Tokyo since 1962. He was an advocate of Metabolism and gained recognition through such projects as his helix city plan (1961). At Expo 1970 in Osaka he was able to put his ideas into practice, as with the Nagakin Kapselturm (Capsule Tower). Uniform, white, living units with circular window openings are fitted to two black, core buildings in a loose arrangement. The units were to be produced in modules so that they could be replaced, in order to remain up-to-date and responsive to change. The architecture that Kurokawa has produced since the 1980s is characterized by the combining of modern with traditional Japanese styles, as with the Museum for Contemporary Art in Hiroshima, Japan (1984–1988). Its façades are made of ceramic, stone and aluminium. Circular and square pillars carry roofs and segments of façade, while a large part of the museum lies underground and is illuminated by large, glass surfaces. The Museum of Modern Art in Wakayama, Japan (1990–1994) occupies the site of the fifteenth century Wakayama castle and adopts the principles of the traditional, bourgeois architecture of Japan with its black and white colors and a shape of the roof and the eaves. There is a striking play of light across a variety of colors. Between 1992 and 1998, and in cooperation with the Jururancang office, he realized the international airport at Kuala Lumpur.

"I consider architecture as part of a space opened to society, rather than as a work of art."

The new Google Office in Moscow is unique and inspiring so many levels as it embraces the local identity and the wants and wishes of the employees. It is immediately welcoming, warm and inviting, relaxed yet focused, comfy and still serious as it provides functional and diverse work environments throughout.

Designed with direct input and collaborative engagement of the local Googlers, the new home brings together, previously separated in two locations, engineering, sales, marketing and corporate services, identifying commonalities, essential needs and the collective spirit of the Googlers.

Throughout the design process, the research analysis and resulting data showed that Moscow employees were strongly aligned with the global identity and values of Google, but they also had a strong desire to reflect these values from a local point of view. They wanted the design to emphasize warmth and family, to reflect the needs of individuals over corporate standards, and they clearly wanted the office to have a strong Russian identity—defined by its rich literature, poetry and natural beauty.

The resulting interior architecture triumphs as it celebrates the culture of the Russian out of finely crafted brickwork mirroring the Cremlin's grand architectural heritage. And in typical Google style, it's playful and fun with areas set aside for video games and ping-pong.

The main office area is light, airy and open, with beautiful views over Old Moscow Centre and the Kremlin. Meeting rooms and informal collaboration areas surround and support the workspace. These are delicately designed at a human, almost residential scale and unified through the use of subtle graphics telling tales from childhood cartoons, movies and fanciful fairytales.

And finally, in the heart of the office there is a special place for Googlers to gather informally in a comfortable, casual setting. It's a large coffee lounge and break area that is super cozy and relaxed. It's a place to unwind, work or quietly converse with coworkers—under a whimsical tree where the mood is magical.

once upon a time
a mouse, a bird and
a sausage entered
a partnership and
house together

in which
to soul
causing
strive to
looking
it if to
too apo
striving to
Mishima

The concept and starting point of the new offices was mainly focused on two purposes, the first one represented in creating a dynamic design, open and sober offices and the second making reference to the vital component of sustainable and eco-friendly to the environment in both the design and construction of the offices and where the commitment of Ael, was absolute.

It was thought from the beginning in the importance of various meeting rooms to meet the demand for client meetings and committees. An open office enjoys with not doubt in a specialty generous thought for each of the departments and outstanding furniture, large work areas for each employees as well as the importance of the filing cabinets that can easily become chairs for informal meetings between two persons. Apart of the meeting rooms, we have rooms for 4 persons to conduct informal meetings internally with contractors or clients.

The "Thinking Room" seeks to create an area where the designers have all the bibliographic tools and materials, where they can do with comfort the process of designing a project, also that space will be used by all the staff, because it also becomes a recreation area so that any employee can spend a pleasant and peaceful time forgetting for a moment of the stress inherent of any office.

SUSTAINABILITY STRATEGIES:

At the entrance was set out a high-void-volume mat to prevent the entrance of particulate material to the working area.

The staircase was made of a metal structure that has a high recycled content. All paints, carpets and furniture used in the project have low emissions of VOC (Volatile Organic Compounds).

The scheme is an open office with few partitions to allow the access of natural light and exterior view. The space counts with natural lighting and a lighting system controlled by occupancy sensors for the closed offices.

The ventilation system is mixed; there are operable windows that allow cross-ventilation and permanent renewal air and a mechanical system of ventilation. The distribution is made by textile ducts. They have less weight and less materials than a conventional ventilation system.

During the course of the work 50% of the waste generated was diverted from landfill to recycling purposes. Over 90% of the equipment has the Energy Star label. In addition to this, the new offices consume 30% less of energy compared to the previous offices when reviewed the invoice of public services. The auditorium with capacity for 120 persons can be divided in 2, 3 or 4 meeting rooms depending on the quantity of people needed. It has sensors that are constantly monitoring the concentration of CO_2. When it reaches 1000 parts per million, it activates and generates an alarm that notifies the users of the room. The chairs and rugs have the Greenguard seal that ensures lower emissions. Dual flush toilets and the urinal consumes 1/8 of water compared with a traditional apparatus. The sinks have push and they have been schedule a shorter cycle to save water.

The savings achieve so far is around 40%.

The building has bikers and the office has showers and dressing rooms to promote bicycle transportation. There are several bikes available for those who need to move closer.

The terrace floor is industrial metal for the ventilation of the basements and power plant. It has a capillary system to capture rainwater and can be self-sufficient for up to 3 months.

The deck and wooden tables have the FSC label.

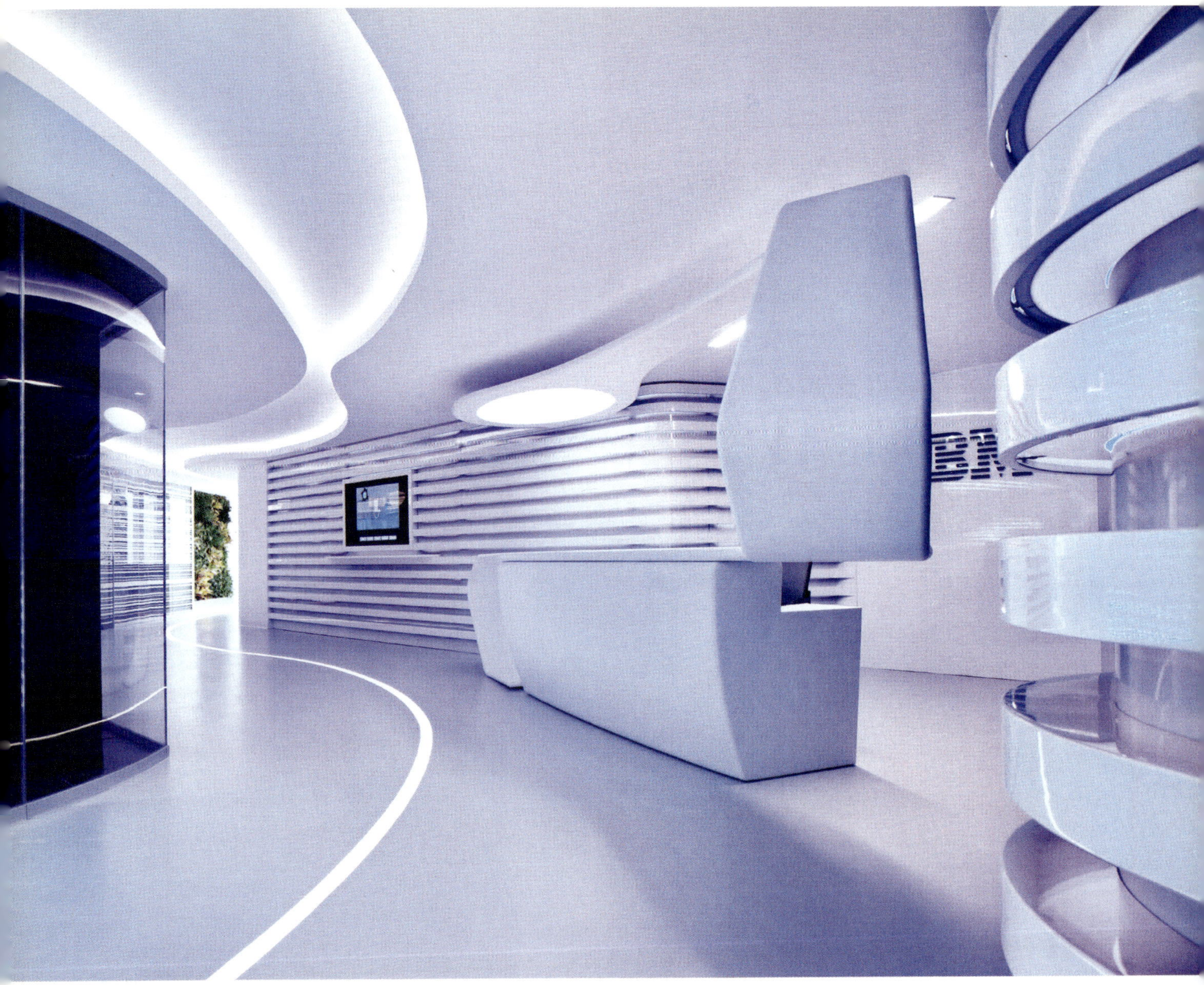

Азия

Екатеринбург
Киев
Новосибирск

← Hovedinngang

Африк

Африка

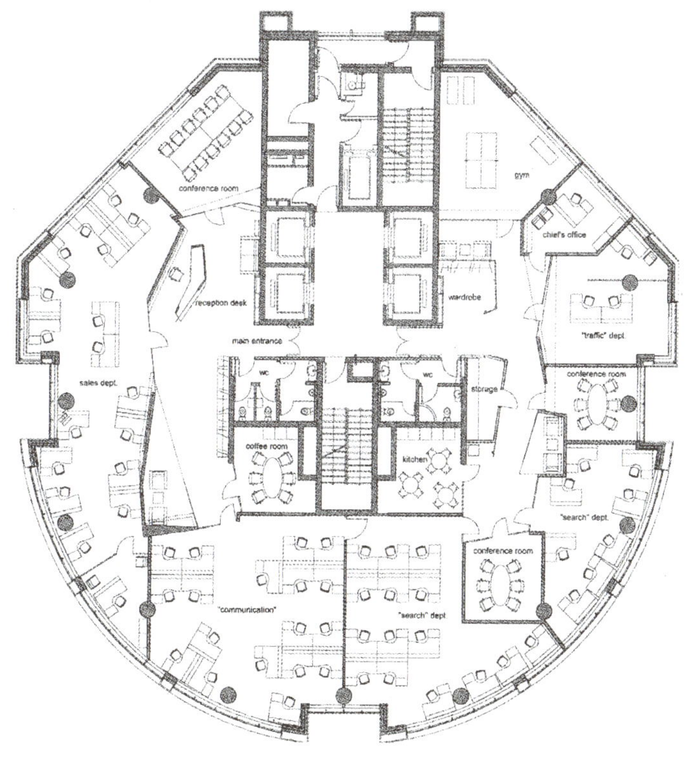

conference room
гур
chief's office
reception desk
wardrobe
"traffic" dept.
main entrance
conference room
wc
wc
storage
sales dept.
coffee room
kitchen
"search" dept.
conference room
"communication"
"search" dept.

KAYAK.COM
Design Company architecture3S
Architect of Record Andrew Cohen Architects
Location Concord, MA
Area 1062 m²
Photographer Greg Premru Photography

2.1 Architectural Form

In middle ages, most of the historical artifacts, art treasures and valuable natural specimens were collected in royal palaces, private mansions, churches, cloisters and universities. In the Renaissance period, the first building designed for art collection, now known as Uffiizi Gallery, is built in Florence, Italy. Britain founded the first public museum – Ashmolean Art and Archaeological Museum. However, from floor layout to elevation, the building remained a palatial style without any functional zones such as display rooms or collection rooms. In 1793, French National Assembly decided to display Louis collection in Louvre to public, marking a palace of old order transferring a museum, an opening to use old buildings as public museums.

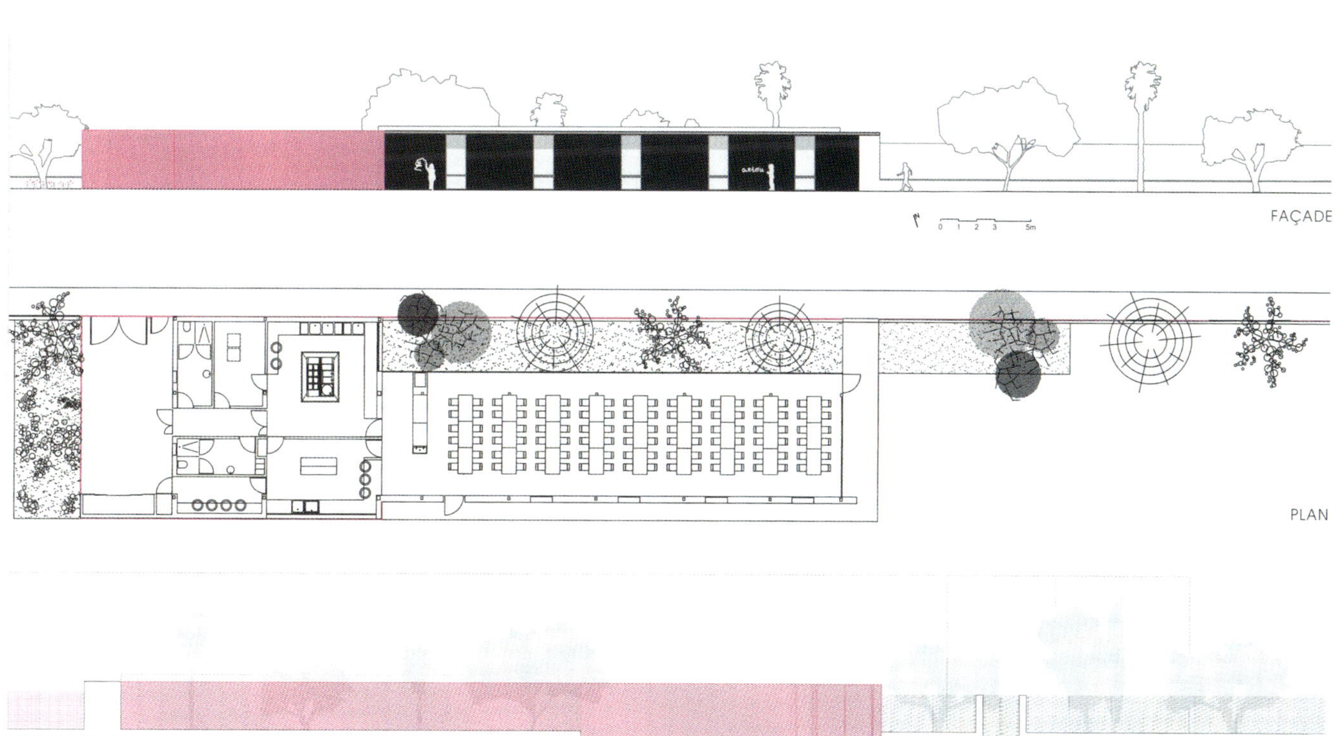
FAÇADE
PLAN
FAÇADE

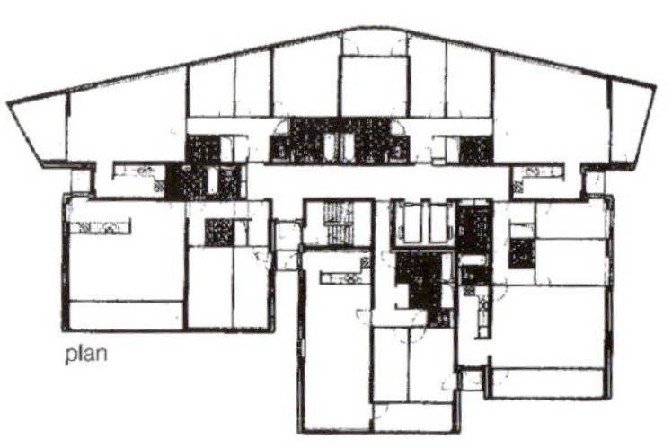

plan

Maastricht is located in the southernmost part of Holland. In a remote corner a little to the west of the city center stands one of Wiel Arets' most representative works, the Maastricht Academy of Arts and Architecture. In the middle of a grove of huge, verdant chestnut trees is an aerial bridge made of reinforced concrete. With the blue sky as a backdrop, the refreshing architectural beauty created by the combination of light, greenery, reinforced concrete, and glass blocks makes for a deeply moving scene.

After passing under the bridge, one finds a square called Herdenkingsplein. Then, looking back, a workshop with glass-block walls in a square grid is visible on the right side, and on the left is the school's old wing, or main building, and adjacent to it, a glass-block auditorium. The bridge, as one might expect, connects the auditorium and the workshop.

Arets, who pays particularly close attention to a building's skin, demonstrates an impressive prowess in his use of glass blocks in this work. The brilliant glass-block walls, which with their square pattern are enclosed in a thin concrete frame, recall the austerity of other glass houses such as Pierre Chareau's Maison de Verre and Tadao Ando's Ishihara Residence.

The use of linear space, something that Arets is deeply fond of, also creates a striking effect. With glass blocks for the floor and ceiling, and reinforced concrete for the walls, the bridge creates a strange feeling of elation rooted in one's sense that the position of the materials would normally be reversed. In the Police Station in Vaals too, two linear spaces stand parallel to each other, creating a floating sensation, with the cantilevers hanging out in space over the slanted lawn area.

Among Arets' other linear-space-based works are the Ceramic Office Building, Groningen Court Building, Delft Theater, Amsterdam Academy for the Arts, the AZL Pension Fund Headquarters, the Police Station in Boxtel, the Lensvelt Office and Factory Building, and Almere Theater. Common to all of these is a long, thin office space, slope, and a corridor – a structure with a heightened sense of perspective.

In looking at Arets' drawings, one finds similarities between his elongated, flowing form of spatial expression and the paths through hyperspace depicted by Zaha Hadid. Arets' expressions, however, are supported by a geometry of rigid right angles. The simple yet pure impression this creates is truly Andoesque, and in light of Ando's use of approach slopes, glass blocks, and reinforced concrete walls, he seems to

"Do you want to know where I found my model? An upright tree; it bears its branches and the branches bear the twigs and the twigs bear the leaves. And each individual part grows harmoniously together, magnificent since God the artist created it."

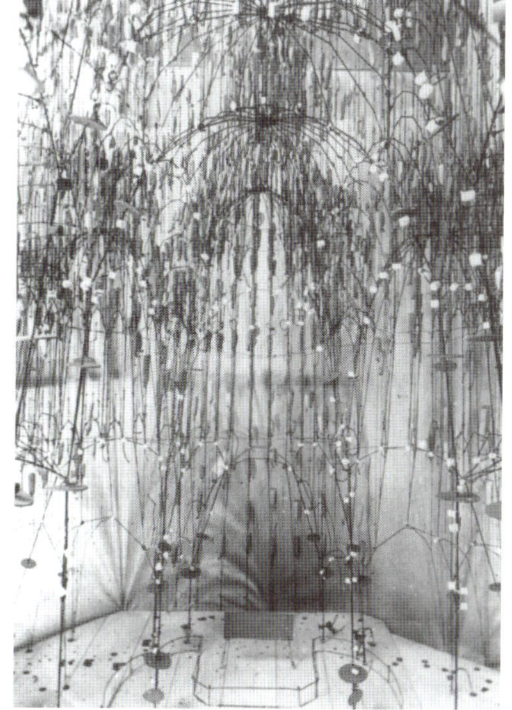

Opposite page: Sagrada Família (Holy Family) cathedral, Barcelona, 1883–1926

Below: Colonia Güell church, Barcelona, Spain, 1898–1915

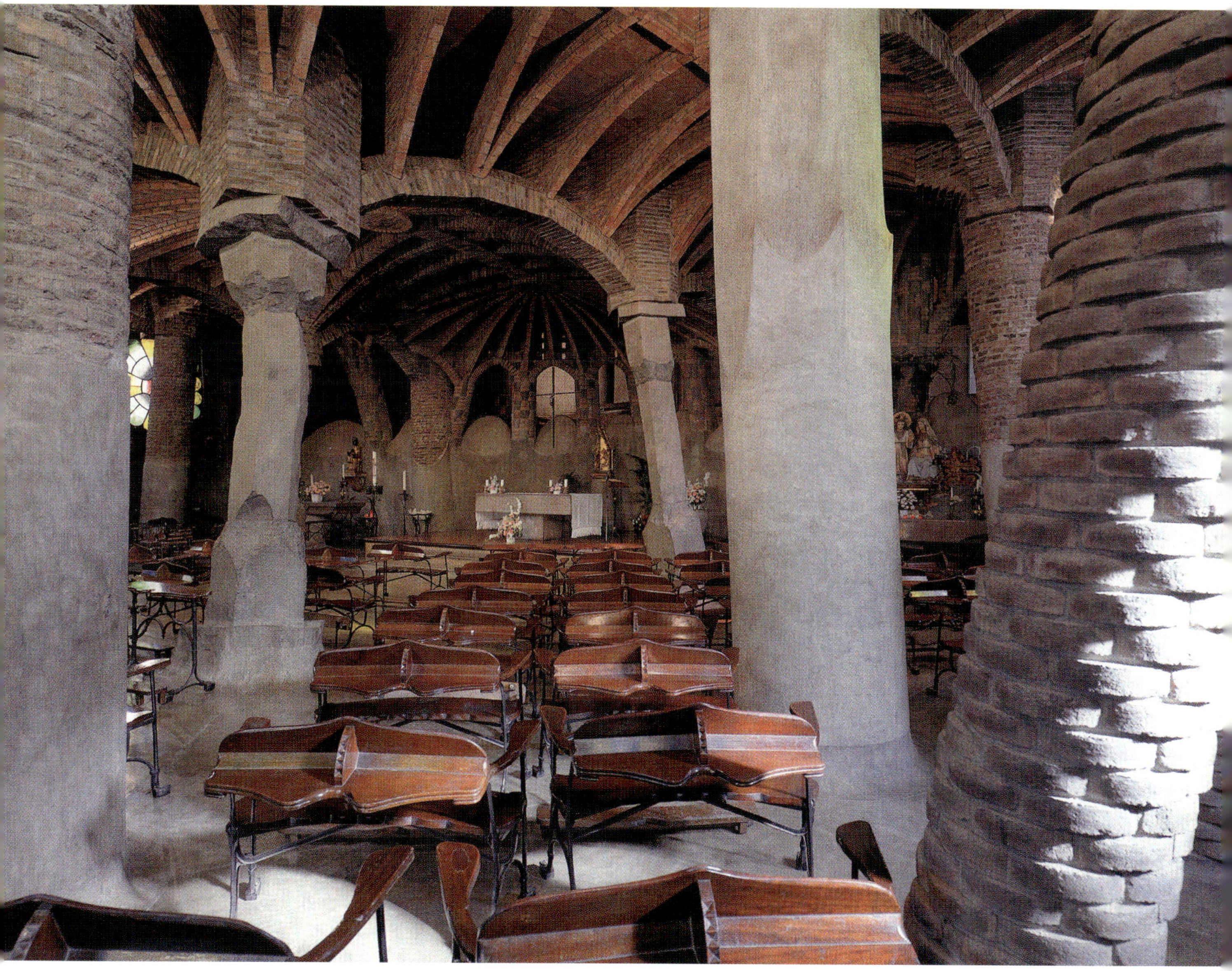

"Functionalism is only justified if it is expanded to cover psychophysical areas too. This is the only way to humanize architecture."

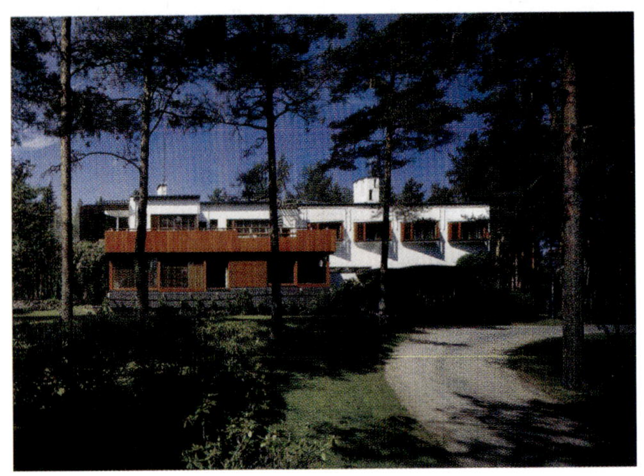

Opposite page and above right: Villa Mairea, Noormarkku, Finland, 1938–1939

Below: Technical University, Espoo, Finland, 1949–1974

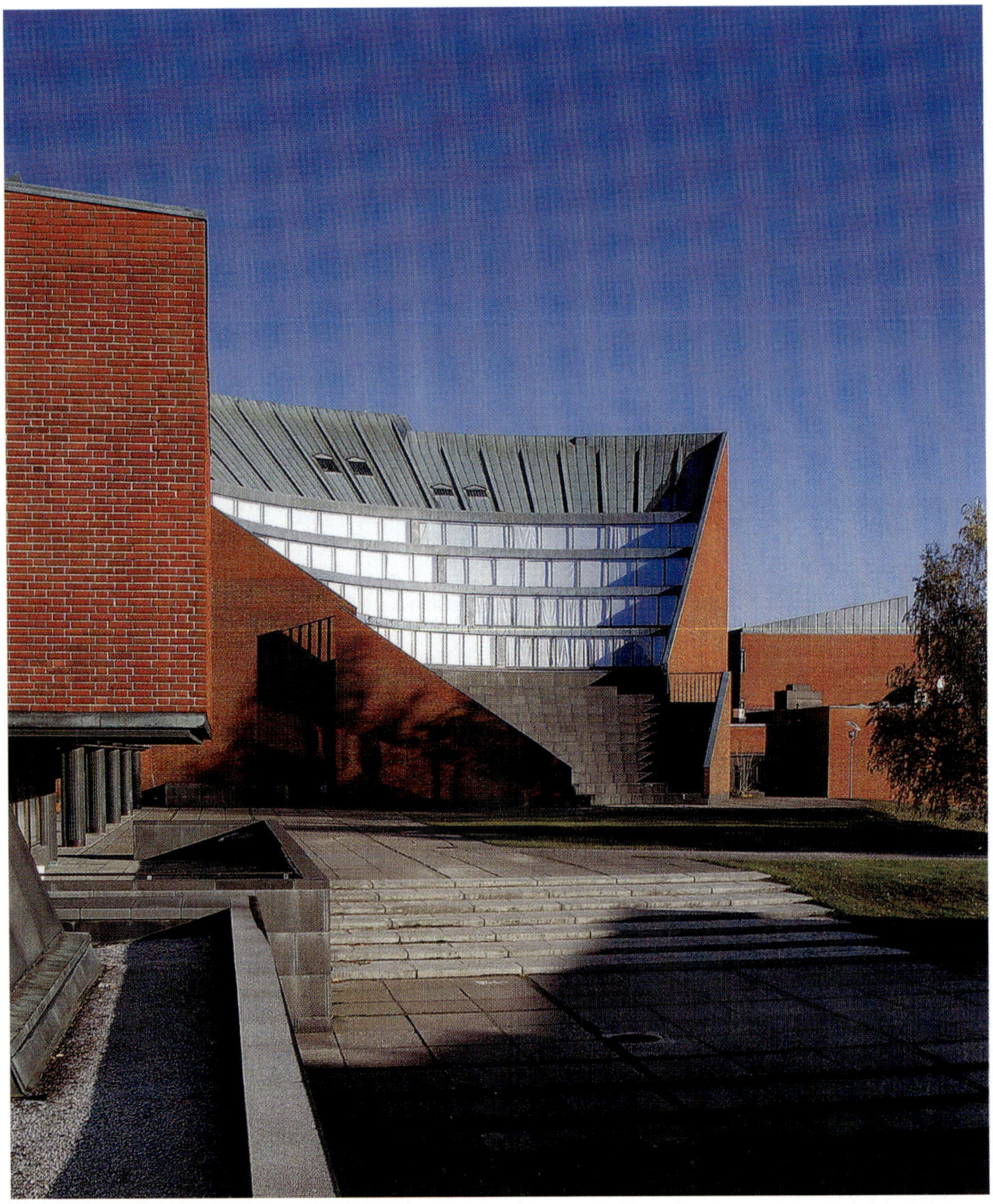

The Canadian architecture world isn't very well known in Japan. The leading light, Arthur Erickson, is remembered for serving as a judge at the Tokyo International Forum Competition, and the Isareli-born Canadian architect Moshe Safdie is fairly famous. Then there is Dan Hanganu, Richard Henriquez, Peter Rose, Patkau Architects, Bruce Kuwabara, and Forsythe and MacAllen, who until just prior to winning the Aomori Northern Style Housing Competition were virtually unknown. Architecturally, Canada is definitely far-removed, but among the most unique of the country's architects is Douglas Cardinal, who those in the know know. Cardinal, an eminent designer in Canada, is also unique for his Native American background. He has said, "I was actually raised in the construction world. I was the oldest of eight brothers and sisters, and was already helping my father make houses and furniture as a child. I think my mother just assumed that I would become an architect." The family lived in a log cabin outside of Red Deer, Alberta, and Cardinal's father worked first as a forest ranger and later as a game warden. Cardinal explains, "We lived way off the beaten track and we studied nature and our own customs."

While in his teens, Cardinal's family moved to the city, and his father began working in the hotel industry. Cardinal and two of his younger brothers devoted themselves to carpentry, helping to build hotels and bungalows, and also making furniture. This lifestyle gave Cardinal an ecological orientation and a pragmatic technique, and helped him develop into an architect. The Canadian Museum of Civilization in Ottawa is a concentration of his way of living and his experiences.

Considered to be one of Cardinal's most representative works, the museum is also the best embodiment of his essence and style. The museum was actually designed as a "habitat" in accordance with the surrounding environment, and the organic design symbolizes human evolution and harmony with the earth. In terms of form, the building's organic curves reflect the rugged nature and varying elevations of the adjacent landscape.

Yet, there are also numerous architectural elements at the museum that perhaps reflect Cardinal's artistic influences. The columns and dome are reminiscent of Alberti, Michelangelo, and Palladio, while the receding steps on each successive level recall the ziggurats of Mesopotamia. And the overhangs and latticework are akin to the Prairie style of Frank

Groningen Court Building / Groningen, The Netherlands
Delft Theater / Delft, The Netherlands
Amsterdam Academy for the Arts / Amsterdam
Police Station in Boxtel   1997 / Boxtel, The Netherlands
photo: ©Kim Zwarts
Cathedral in Ghana / Cape Coast, Ghana
Lensvelt Office & Factory Building
1999 / Breda, The Netherlands
photo: ©Kim Zwarts
Almere Theater / Almere, The Netherlands
Utrecht University Library   2004 / Utrecht, The Netherlands
photo: ©Synectics

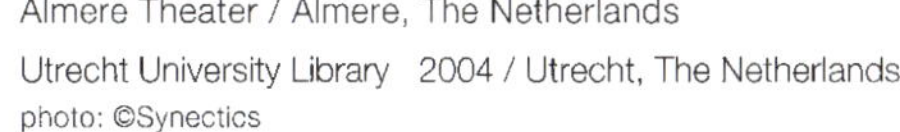

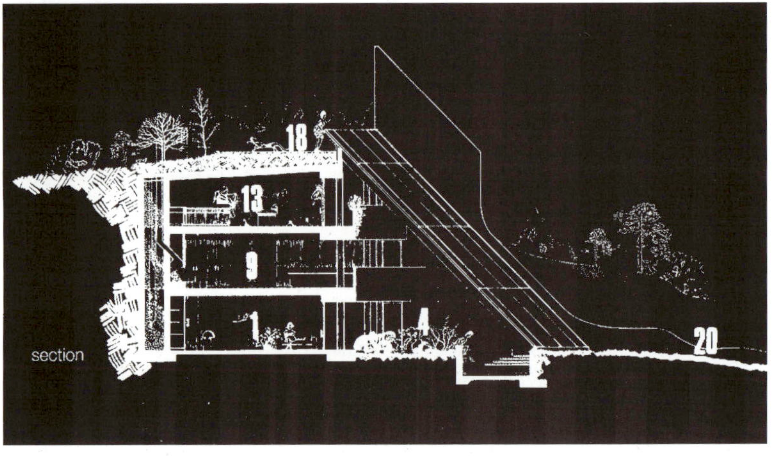

Canadian Museum of Civilization   1989 / Quebec, Canada
Cardinal Studio & Residence   1982 / Alberta, Canada

plar

Federal Agency for the Environment   2005 / Dessau, Germany
photo: ©Jan Bitter

Experimental Factory in Magdeburg   2001 / Magdeburg, Germany
photo: ©Gerrit Engel

Hennigsdorf Town Hall   2003 / Hennigsdorf, Germany
photo: ©Bitter+Bredt

Zumzobel Staff   1999 / Berlin
photo: ©Bitter+Bredt

Pharmacological Research Laboratories in Biberach   2002 / Biberach, Germany
photo: ©Jan Bitter

The British Council in Germany   2000 / Berlin
photo: ©Bitter+Bredt

Scheduled for completion in 2007 is the Performers House in Denmark; in 2008, the Amazon Court in the Czech Republic; in 2009, Thor Heyerdahl College in Norway; and in 2011, Aberdeen University Library in Scotland. From this list, it's clear that the firm's outstanding brand of design has been recognized throughout Europe.

Perhaps the key to the high-quality work SHL has created lies in the excellent organizational structure of the firm. In addition to SHL's specialty in building design, the firm includes four separate divisions which support its architectural projects: a special project division, a furniture and fittings division, a landscape design division, and a graphic design (communications) division.

Founded in 1986 by Morten Schmidt, Bjarne Hammer, and John Lassen, some 20 years later, SHL has grown into one of the largest architectural firms in Denmark, with a staff of over 100 people. Following the later addition of Kim Holst Jensen, and more recently Morten Holm, the firm now operates on a five-partner structure. And despite his long career with SHL, at 52, Lassen is still a young man. After graduating form the Architecture School at the University of Aarhus, the founding members set up SHL's main office in Aarhus and a branch in Copenhagen. The architects, whose modus operandi calls for assembling a good team to create design, say, "Architecture today is extremely complicated and it's very difficult for only one architect to see to the design, engineering and management of a project. To realize quality architecture, a team of strong individuals is necessary. What's most important is how well these individuals can join forces to make a solid collaborative unit." This knowing outlook ranks with Renzo Piano's admirable observation, "The creation of architecture isn't the result of the efforts of a single person, but the gift of teamwork."

SNOHETTA

NORWAY

Firm founded in Oslo in 1987. Current principals are Craig Dykers (born in Frankfurt in 1961; American nationality), who graduated from the University of Texas in 1985, and Kjetil Thorsen (born in Norway in 1958), who graduated from the Technical University of Graz in 1985. Snohetta is both an architecture and a landscape-architecture firm.

ROTO ARCHITECTS

USA

Firm founded by Michael Rotondi, who was born in Los Angeles in 1949. Graduated from the Southern California Institute of Architecture (SCI-Arc) in 1973, and has taught at the same institution since 1976. In 1975, established Morphosis with Thom Mayne. The two continued their partnership until 1991, when Rotondi founded Roto Architects. In addition to being one of the "founder-students" of SCI-Arc, also served as director of the school for ten years from 1987 to 1997.

town square to the interior, Rossi arrived at an excellent work which, in his familiar style, established a close link with the city.

Rossi's work is called "analogical architecture." For example, he sometimes referred to a painting of Venice by the 18th-century Italian painter Canaletto which hangs in the National Gallery in Parma. In the painting one finds further references to three works by Palladio: "Design for the Rialto Bridge," "Basilica," and "Palazzo Chiericati." The subjects of the latter two are Vicenza rather than Venice. Therefore, Rossi explained, the city in the picture is an analogy for Venice, and this is what inspired his analogical architecture.

The foundations of Rossi's analogical architecture are the old barn, corridor, factory, silo, and warehouse that slumbered in the depths of his memory. These surreal and plain forms that seem to have been frozen in time served as an analogy for the rational architecture of the Gallaratese Housing Complex. The conjoined walls and columns and the long corridors give the building a terse, metaphysical air.

In the 80s, Rossi began work on a housing complex as part of the International Building Exposition (IBA) in Berlin, and at the same time, in Italy, he developed large works such as the Centro Torri Shopping Center and Casa Aurora. Following the completion of the Il Palazzo, Rossi also created a series of works in Japan. These eight buildings include the Ambiente Showroom, Asaba Design Office, and Mojiko Hotel. He also pursued a successful collaboration with the noted Japanese designer Shigeru Uchida.

In 1990, Rossi was awarded the Pritzker Prize, which is considered to be the Nobel Prize of the architecture world. Next, he expanded his activities around the globe. In particular, Rossi's many works in the U.S., including the Celebration Building, the Pocono Pines House, the ABC Building, and the Scholastic Building. In addition, as seen in works such as the Bonnefanten Museum in Holland and the Center for Contemporary Art in Vassiviere in France, Rossi's work became elegant and beautiful.

However, in 1997, Rossi was again involved in a car accident. This time the man whose life had been so greatly altered by a crash lost his life. I still have fond memories of drinking _sake_ with him at Il Palazzo, and appearing on TV together at the Rossi exhibition in Ginza. His death at 66 occurred far too soon and left many sad people behind as the master architect flitted across the skies of Lombardia.

Cuban people had gone through in the revolution. Like a pair of fragmented glasses, the studio consists of two spaces and with its centripetal design, the work exudes energy. In 1960, after completing these two works, Porro moved to France. From then until 1992, he served as an instructor at a variety of different schools. In the meantime, he became a sculptor, and also developed his skills as a painter. This led him to regularly include his own sculptures and paintings in his architectural works. For example, in 1975, Porro added "Sculpture of Mouths," depicting a strange human form, to the Office and Art Center, his first European work. And in the courtyard of the School of Plastic Arts in Havana, he created the aforementioned papaya sculpture as a symbol of the female sexual organ.

In 1986, Porro began working with the architect Renaud de la Noue, and together the two designed 30 Residential Dwellings in Stains as well as the square on the site. From this point on, Porro suddenly became more and more active in France. Among his works were College de Cergy-le-Haut, which referenced the traditional French castle Ch_teaux de la Loire; College Elsa Triolet, which made use of a dove motif; and the La Courneuve Dwellings, notable for its sidewalls, which resemble human ears. In addition, Porro designed College Fabien, a hermaphroditic example of architecture that combined both male and female components, and the Barracks of the Republican Security Force in Velizy, which was inspired by Paulo Uccello's painting "The Battle of San Romano." One after another, all over France in particular, Porro produced organic and androgynous architectural masterworks. With his experiences in Cuba as a foundation, he had become especially deft at designing schools, hence the large number of educational facilities.

Born into a wealthy family and a fervent reader since childhood, Porro came to love Plato, Dante, Thomas Mann, Martin Heidegger and Marcel Proust in his later years. In architecture, he was hugely influenced by Michelangelo, Borromini, Frank Lloyd Wright, and Erik Gunnar Asplund. From his Cuban period, when he deviated from the logic of modern architecture, to his contemporary work, which has been called heretical for its mixture of expressionistic, epicurean, organic, and hermaphroditic elements, Porro has consistently received acclaim for his idiosyncratic exteriors and the stark contrast created by his rich, warm interiors.

Karikatur Museum   2001 / Krems, Austria
photo: ©Martin Wawra
KITA-Kindergarden   1999 / Berlin
Messe Wien   2003 / Vienna
photo: ©Gisela Erlacher
ORF Station in St. Polten   1994 / St. Polten, Austria
EVN Forum   1993 / Maria Enzersdolf, Austria
Phosphate Elimination Plant, Berlin-Tegel   1985 / Berlin
photo: ©Synectics
Austria Tower

BEAUTIFUL PEOPLE
THE WAR ZONE
"MIFUNE"
CENSORSHIP

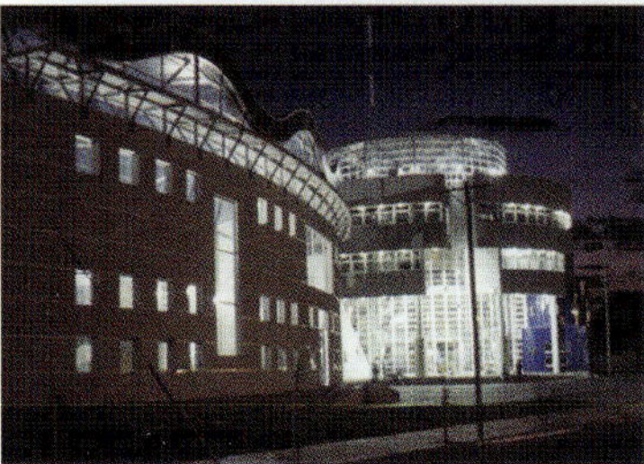

Spain Plaza 1969 / Cordoba, Argentina
CPC Monsenor Pablo Cabrera 1999 / Cordoba, Argentina
Underground Passage between Two Museums in Cordoba
2006 / Cordoba, Argentina

photos : Courtesy of the Architect.

planning involving the construction of a new city in place of an already existing one. Moreover, in works like Spain Plaza and Italia Plaza, Roca, a student of Louis Kahn, achieved a special charm by filling the sites with swinging, heavy geometrical objects of the kind that Kahn used in his non-residential works. La Florida Park, in La Paz, is one of Roca's most poetic and moving works. On a long, thin site along a river stands a roofed colonnade that undulates like a snake. Intersecting it at many points are reinforced concrete walls, which create a wondrous and tension-filled urban space as ambiguous fragments arranged in a line.

The surfaces of the numerous walls are painted brilliant colors that resemble those of _awayo_, the traditional Indio belts worn by women. They are also similar to the pink wall surfaces used by Luis Barragan in San Cristobal, and the two works both have large wall openings. Barragan once wrote, "The garden is poetic, mysterious, enchanting, serene, and enjoyable." In the shade of the trees that is cast across the colorful, Latin-style walls in the park, one detects the "spatial poetry" that underlies Barragan's work.

Miguel Angel Roca was a precocious child born to a Cordoba architect. Except for an interest in drawing, the young Roca showed no inclinations toward architecture. But he went on to do post-graduate work at the University of Pennsylvania with Louis Kahn, and later worked at Kahn's firm. Roca sees architecture as something that is much more than a simple method of aesthetic expression and serves as a sociopolitical medium that provides physical shelter and structures an environment. In addition to the poetic flavor of his work, one finds political overtones. As a result, some have come to refer to Roca's work as "poetic-political" architecture.

Along with the recently completed House at Calamuchita, at the end of 2005, Roca finished large, urban works such as Underground Museum Passage in Cordoba and Renovation for Corrientes Street.

La Florida Park   1990 / La Paz, Bolivia
Faculty of Law at the National University of Cordoba   2001 / Cordoba, Argentina

plan

ALDO ROSSI

ITALY

Born in Milan in 1931. Graduated from Politecnico di Milano in 1959, and opened his own firm the same year. Taught at the University Iuav of Venice and Politecnico di Milano. Member of the editorial board at the architectural journal _Casabella Continuita_ from 1961 to 1964. Beginning in 1971, taught for several years in Switzerland. Became a professor at University Iuav of Venice in 1975. Awarded the Pritzker Prize in 1990. Died in a car accident in 1997.

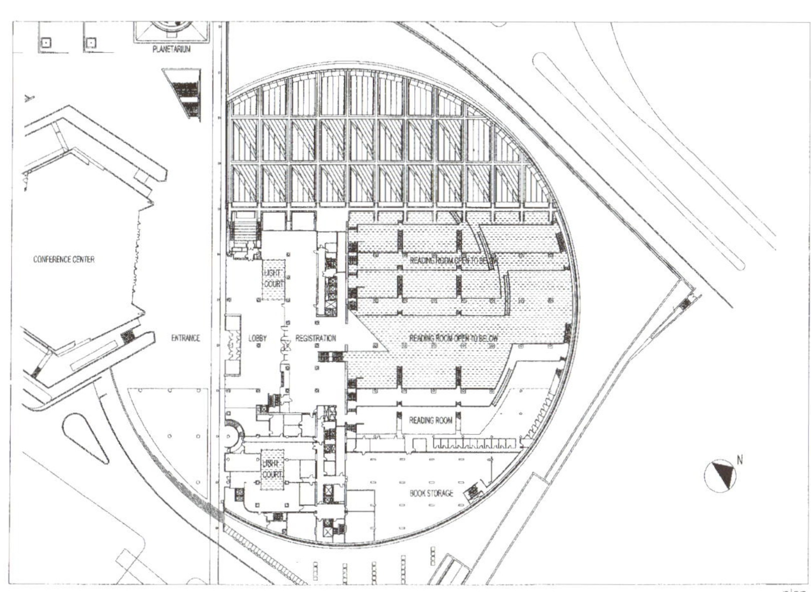

KHIB-National Academy of the Arts in Bergen / Bergen, Norway

Alexandria Library 2002 / Alexandria, Egypt
photo: ©Gerald Zugman

plan

Alte Mainzer Gasse
...tergasse

Below: Notre-Dame-du-Haut
church of pilgrimage,
Ronchamp, France, 1951–1955

Opposite page: Unité
d'habitation, Marseilles, France,
1945–1952

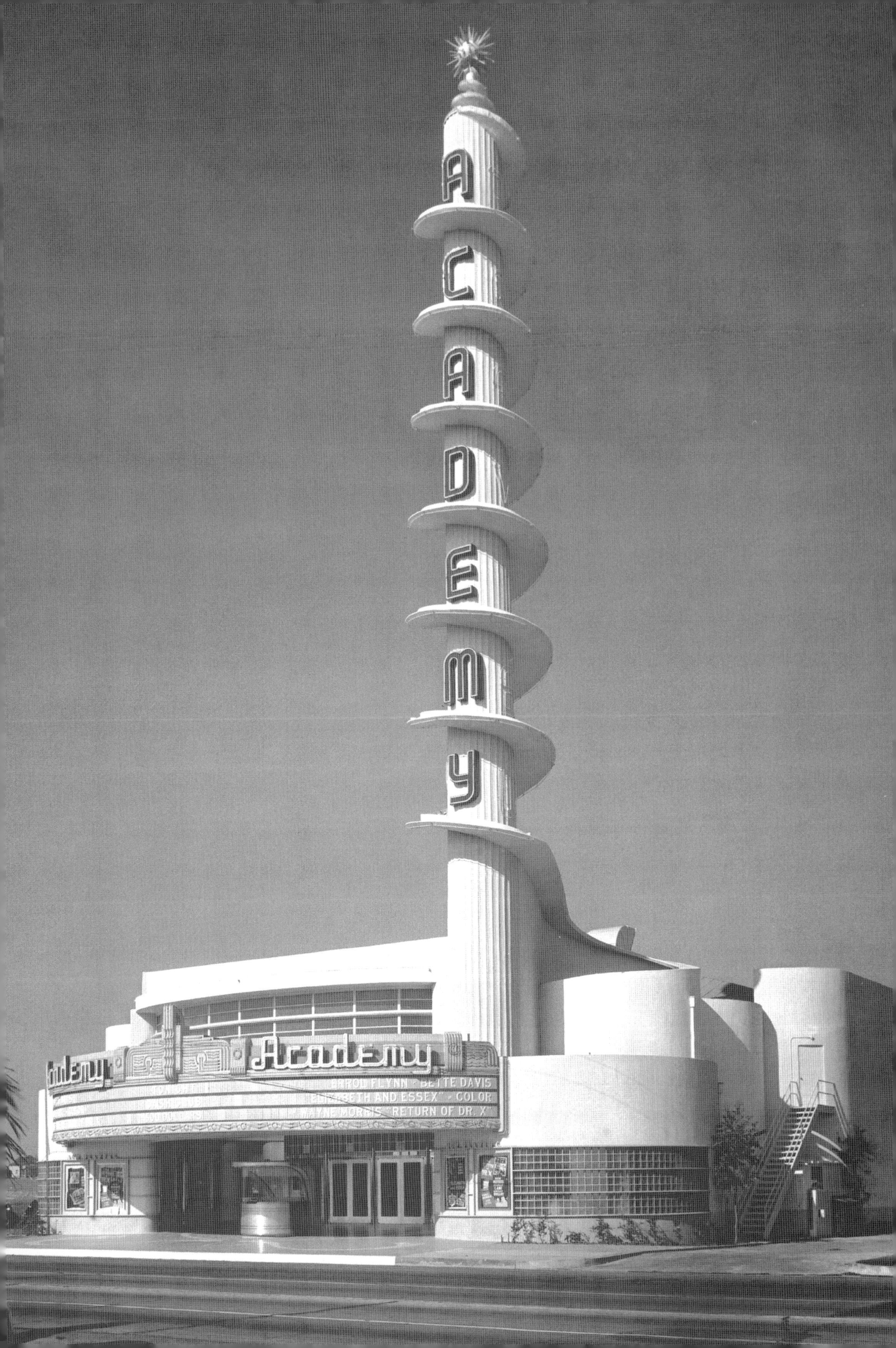

ACADEMY
Academy
ERROL FLYNN · BETTE DAVIS
ELIZABETH AND ESSEX · COLOR
WAYNE MORRIS · RETURN OF DR. X

Opposite page: Arango house,
Acapulco, Mexico, 1973

Above: Garcia house,
Los Angeles, California, 1962

Below: Malin house,
"Chemosphere", Los Angeles,
California, 1960

Jahn Helmut

After graduating from the Technische Hochschule in Munich, Helmut Jahn (born in 1940), trained under Peter C. Seidlein until 1966, gaining additional experience in 1965 and 1966 at the Illinois Institute of Technology in Chicago. In 1967, he joined C. F. Murphy Associates as assistant to Gene Summers. After Summers left the firm, Jahn took over his position as head of the Planning and Design Department. In 1981, he became a partner in the firm, renamed Murphy/Jahn, and in 1982 was appointed its President, before taking over sole responsibility for its management in 1983. One of the firm's projects was the completion in 1975 of the Crosby Kemper Memorial Arena in Kansas City, Missouri, designed to accommodate 1,800 people. This was completed within a very short space of time and on a modest budget. Its roof is held in suspension by three giant trusses and its façades are clad in metal panels. The athletics facility at St. Mary's College in Notre Dame, Indiana (1977), is day-lit from the sides, giving the space, with its red supports and blue and yellow pipes, a cheerful appearance. The State of Illinois Building in Chicago, a 17-story complex with a useable area of 103,500 m², was built between 1979 and 1985 on a quadrant ground plan. The glass membrane is pierced by a glass cylinder, which creates a circular public area and atrium in the interior. Since 1981, Jahn has been a professor of architecture at Harvard University. In the early 1980s, he designed an extension to the Stock Exchange building (1979–1982), the skyscraper at One South Wacker Drive (1980–1982) and the Northwestern Atrium Center (1984–1987) in Chicago. Other commissions include the Frankfurt Messeturm (Exhibition Tower; 1985–1991) with its pointed spire, as well as the Sony Center in Berlin (1993–2000). This giant complex, consisting of offices, shops, theaters and hotel, is arranged around the Sony Forum and roofed with a lightweight construction of glass, cables and sailcloth. Shade-dispensing membrane roofs also stretch across the glass covering of Bangkok airport (1995–2004).

Judge Bernard

Bernard Judge (born in 1931) attended the École des Beaux-Arts in Paris and the University of Southern California, from where he graduated in 1960. His studio, the "Environment Systems Group", has been in existence since 1965 and is involved above all in the design and construction of private homes, hotels and holiday resorts. His "Dome House" in Los Angeles, a spherical residence with several levels and an exterior divided into small triangles, was built in 1960. The basic structure is a geodetic cupola by Buckminster Fuller that Judge, as something of an experiment, developed into a private home. The experiment was demolished in 1970. In 1971, in the interests of sustainable planning and building, he developed a master plan for Marlon Brando's Tetiaroa Atoll in Polynesia. His own home, built in 1975, was also conceived as a model house. Specially designed for hillside situations, it is based on a pillar construction that requires little excavation. Four, concrete cassette panels form the base and carry the four, fifteen-meter-high main supports for the wood frame construction. The non-load-bearing interior walls can be moved as required.

Kuma Kengo

Kengo Kuma (born in 1954) studied architecture at the University of Tokyo. For two years in the middle of the 1980s, he was a Visiting Fellow at the University of Columbia in New York, before founding the Spatial Design Studio in Tokyo in 1987. In 1991, he formed the office Kengo Kuma

*"I want to 'erase architecture'.
I have always wanted to do so."*

& Associates. That year he also created the M2 multi-purpose building in Tokyo for the car producers, Mazda. With its overtones of portals, immensely high "Ionic" pillar, inner courtyard and lift it is very different from the projects that followed, such as the Noh stage, built on a wooded area in Toyama, Japan (1996), and the River/Filter restaurant in Fukushima, Japan (1996). The contours of this latter building are dissolved by layers of wooden lattices, which are made up of rows of slats of varying breadth and thickness set close together. Further designs produced by Kuma include the Kitakami canal museum in Miyagi, Japan (1999), which is fused into a hillside and whose architectural form is subjugated to its surroundings, and a school in Kanagawa, Japan (1999), whose walls of vertically arranged slats enable those inside to look out at nature. He completed the Stone museum in 2000 in Nasu, Japan, in which the building materials form part of the exhibits. For the Lotus house in Kamakura, Japan (2003–2005) he used a mixture of materials, including bamboo, glass, stone, plastic and metal. The façade comprises a mesh made up of thin, stainless steel struts with white, travertine plates hung on them.

Opposite page: Government
quarters, Dacca, Bangladesh,
1962–1983

Below: Student accommodation,
Indian Institute of Management,
Ahmedabad, India, 1962–1974

Isozaki Arata

After studying under Kenzo Tange at the University of Tokyo (until 1954), Arata Isozaki (born in 1931) continued accumulating practical experience working in Tange's practice. In 1963, he opened his own office in Tokyo, while continuing to work for Tange on various projects, such as designing the Festival Plaza at Expo '70 in Osaka, Japan (1966-1970). The first art museum to be built by Isozaki was the Museum of Modern Art in Takasaki, Japan (1971-1974), a basic rectangular structure consisting of 12 meter cubes. Angled diagonally to these is a projecting wing, surrounded by water, with an open area at first-floor level. Among Isozaki's best-known works

are the Tsukuba Center Building in Tsukuba (1979-1983), the Art Tower, Mito (1986-1990), an aluminium tower in the form of a double helix, the B-con Plaza in Oita (1991-1995) and the Museum of Contemporary Art, Nagi (1991-1994). Combining wood, stone and metal, this museum comprises three parts: a cylinder, a semi-circular area and a long flat building, originally designed to house the exhibitions of three different artists. Isozaki also carried out several major projects abroad, including the Museum of Contemporary Art in Los Angeles (1981-1986), the Centre for Japanese Art and Technology in Krakow, Poland (1990-1994) and the Palau Sant Jordi (1983-1992)

for the 1992 Olympic Games in Barcelona. The Higashi Shizuoka Cultural Complex in Shizuoka, Japan (1993-1998), with its high, sloping side walls and rounded rear section is a mixture of Japanese tradition and modern art. Between 1993 and 1999, Isozaki designed the Center of Science and Industry in Columbus, Ohio, as well as the Berlin Volksbank building (1993-1998) near Potsdamer Platz. Since 1964, he has lectured at Tokyo University, with guest professorships at Harvard University, Yale, UCLA among others.

HPP
(Hentrich-Petschnigg & Partner)

Administrative building for
Phoenix-Rheinrohr AG, Düsseldorf, Germany, 1956–1960

Helmut Hentrich (1905–2001) studied in Vienna
and at the Technische Hochschule in Berlin. Before establishing his own practice in Düsseldorf
in 1933, he lived for two years in Paris and New
York, where he worked for Norman Bel Geddes.
Between 1935 and 1953, he maintained a partnership with Hans Heuser, and then with Hubert
Petschnigg. They realized the high-rise building
for Phoenix-Rheinrohr AG in Düsseldorf (1956–
1960), consisting of three, offset disks at varying
heights. The Standard Bank building in Johannes-
burg, South Africa (1965–1970), a 139-meter-high
suspended structure, was built in collaboration
with Ove Arup.

"Architects must be able to speak the language of their time because architecture is a public art, it is an art for people. Paradoxically, it is only then that architecture can last forever, only then that it is more than just a creation for the moment." Jacques Herzog

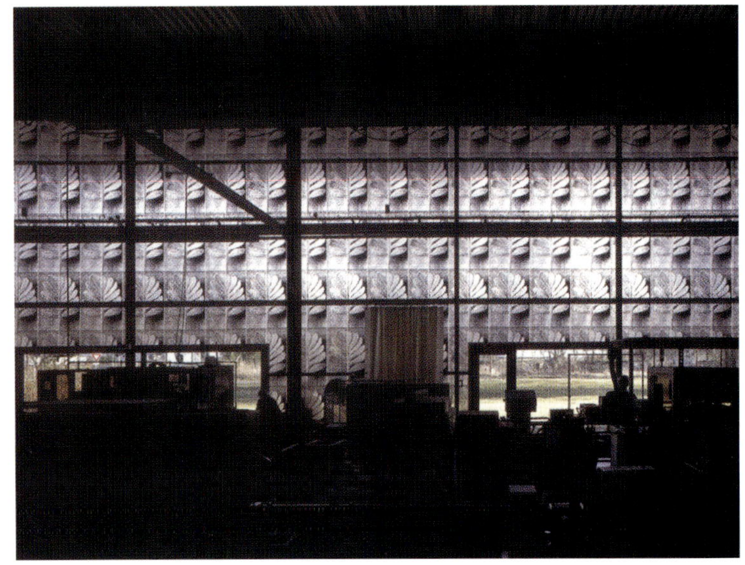

Above: Ricola production and storage building, Mulhouse-Brunstatt, France, 1992–1993

Below: Prada Aoyama Epicentre, Tokyo, Japan, 2000–2003

Opposite page: Allianz Arena, Munich, Germany, 2001–2005

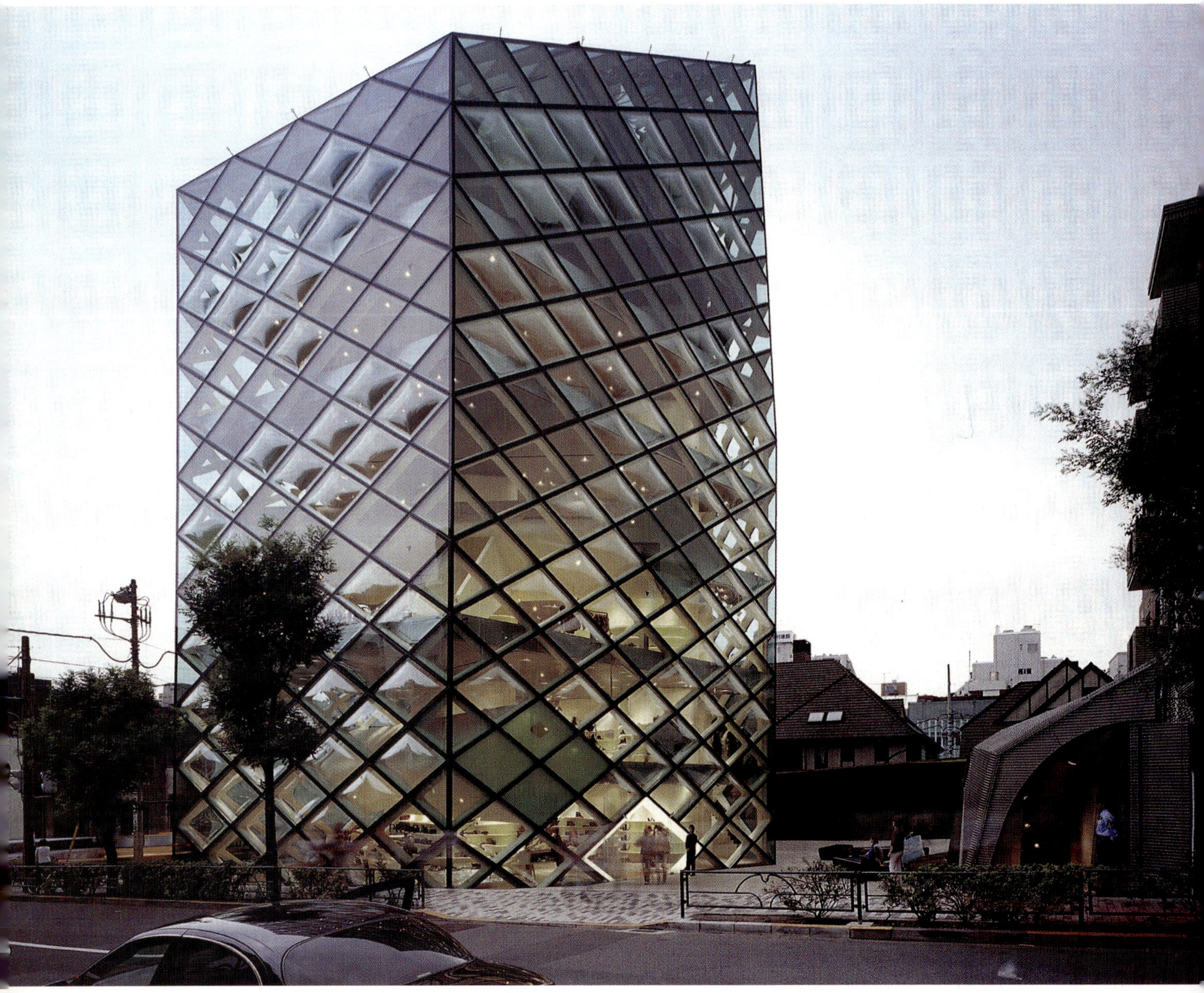

Holl Steven

Steven Holl (born in 1947) studied at the University of Washington and the Architectural Association in London until 1976 before opening his own studio in New York. After teaching at the universities in Washington and Pennsylvania, Holl became a professor at Columbia University, New York in 1981. His early works include the Hybrid Building in Seaside, Florida (1988), an experimental residential construction in Fukuoka, Japan (1989–1991) and the Stretto house in Dallas, Texas (1990–1992). Playing with the different ways light falls in and on a building is typical of his work; it is used to particularly striking effect in the offices of the Shaw real estate agency in New York (1991–1992). Cleverly positioned, pigmented surfaces color the light that penetrates the economical openings in the cube-shaped entrance area. In 1994, Holl collaborated with artist Vito Acconci for the design of the window- and door-less façade of the gallery Art & Architecture in New York (1992–1993). The façade is opened by means of differently sized panels that tilt vertically and horizontally. He also created the chapel of St. Ignatius in Seattle, Washington (1995–1997) and the Kiasma Museum of Contemporary Art in Helsinki (1993–1998), an unusual design resulting from the penetration of two structures, one strictly orthogonal and one with a rounded outer wall. The offices on Sarphatisstraat in Amsterdam (1996–2000) were added to an existing brick building that dates back to the nineteenth century. The façade is clad in perforated copper and presents a charming contrast to the old building. Inside, the perforated pattern is continued in the design of the walls. Window openings are partly covered by the permeable copper sheet; some of the windows have colored panes. The Bellevue Art Museum in Washington was realized between 1997 and 2001. Here, Holl has designed a building with exhibition areas on three levels. The bright red of the façades is interrupted by deep incisions. Although the structure of the student halls of residence at Simmons Hall in Cambridge, Massachusetts (1999–2002), the result of a long design process and called the "sponge", was retained visually, the functional core was vitiated by various safety requirements; this meant that hardly any of the areas required for informal encounters were left.

"Ideas, not shapes or styles, are the most promising heritage of 20th-century architecture."

Hara Hiroshi

Umeda Sky City, Osaka, Japan, 1988–1993

Hiroshi Hara (born in 1936) studied at Tokyo University, where he subsequently taught as a professor of architecture. He started to collaborate with Atelier Phi in 1970. He designed the Hara house in Machida, Japan (1973–1974), the Yamato International building in Tokyo (1985–1986) and the Umeda skyscraper in Osaka (1988–1993) clad in mirrored glass, consisting of two linked towers with a 150-meter-high "hanging garden". The end of the 1990s saw the completion of the railway station in Kyoto (1991–1997) and the Ito house in Chijiwa, Japan (1997–1998). The latter consists of strictly geometrically designed, separate structural units. The tower-like areas for parents and children, built on a square plan, consist of a concrete foundation with a glazed story immediately above and a timber-clad upper section. The rounded, shiny, metallic structure of the Sapporo Dome in Hokkaido, Japan (2001) lies in the valley like a flattened egg. The size of the 42,000-capacity, all-weather sports stadium is only apparent when you approach it. The Orimoto House in Uchiko-Ko, Ehime, Japan (2003) consists of a main building constructed on a quarter-circle ground plan, with a teahouse adjoining an inner courtyard. Both inner façades are glazed, creating extra visual dimensions.

"Designing houses on the moon is an interesting experiment in thought for me."

Graft

Bottom: Hotel Q!, Berlin, Germany, 2002–2004

Below right: Dental clinic KU64, Berlin, Germany, 2004

Lars Krückeberg (born in 1967), Wolfram Putz (born in 1968) and Thomas Willemeit (born in 1968) studied at the Technische Universität (Technical University) in Brunswick. Krückeberg then went on to study in Florence and Los Angeles, while Putz went to Salt Lake City and Los Angeles and Willemeit to Dessau and Vienna. All three architects have worked together in Los Angeles since 1998, under the collective name of Graft. They also now have subsidiary offices in Berlin and Beijing. Up until now they have been mainly concerned with interior design. For the Hotel Q! in Berlin (2002–2004), they designed rooms with walls and fittings in flowing forms. These free forms, opening up an alternative world to that of the conventional portfolio dominated by right angles, were also used in the dental clinic KU64 in Berlin (2004) and the "Fix" restaurant of the Bellagio casino in Las Vegas (2003–2004). A foyer in warm yellow tones greets the client in the dental clinic. Computer-generated images are beamed onto counters, floors and ceilings. With its ceilings hung in the form of waves and its reddish-brown color scheme, the "Fix" produces a cave-like effect. An attic flat in Berlin (2003–2004) was built atop a turn-of-the-century house. The curved surface that extends throughout the whole floor conceals the functional rooms and leads the way through the U-shaped floor plan. Recesses and wall units, as well as built-in cupboards and furniture, define the remaining external walls next to the all-glass façade, creating a futuristic mood with their extraordinary, rounded forms.

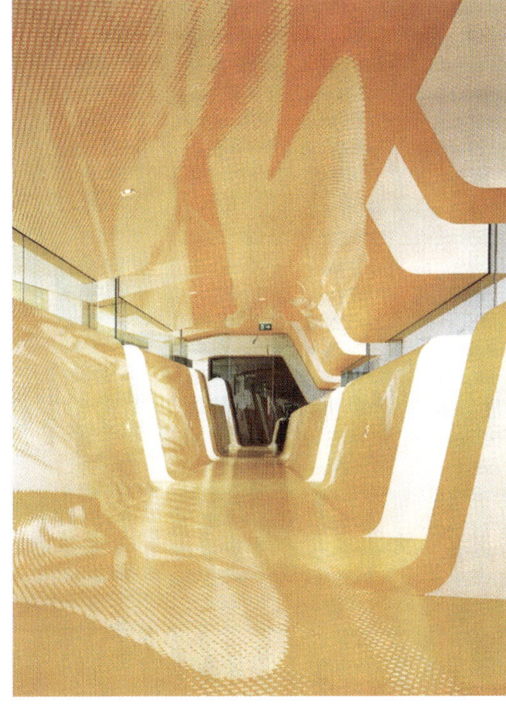

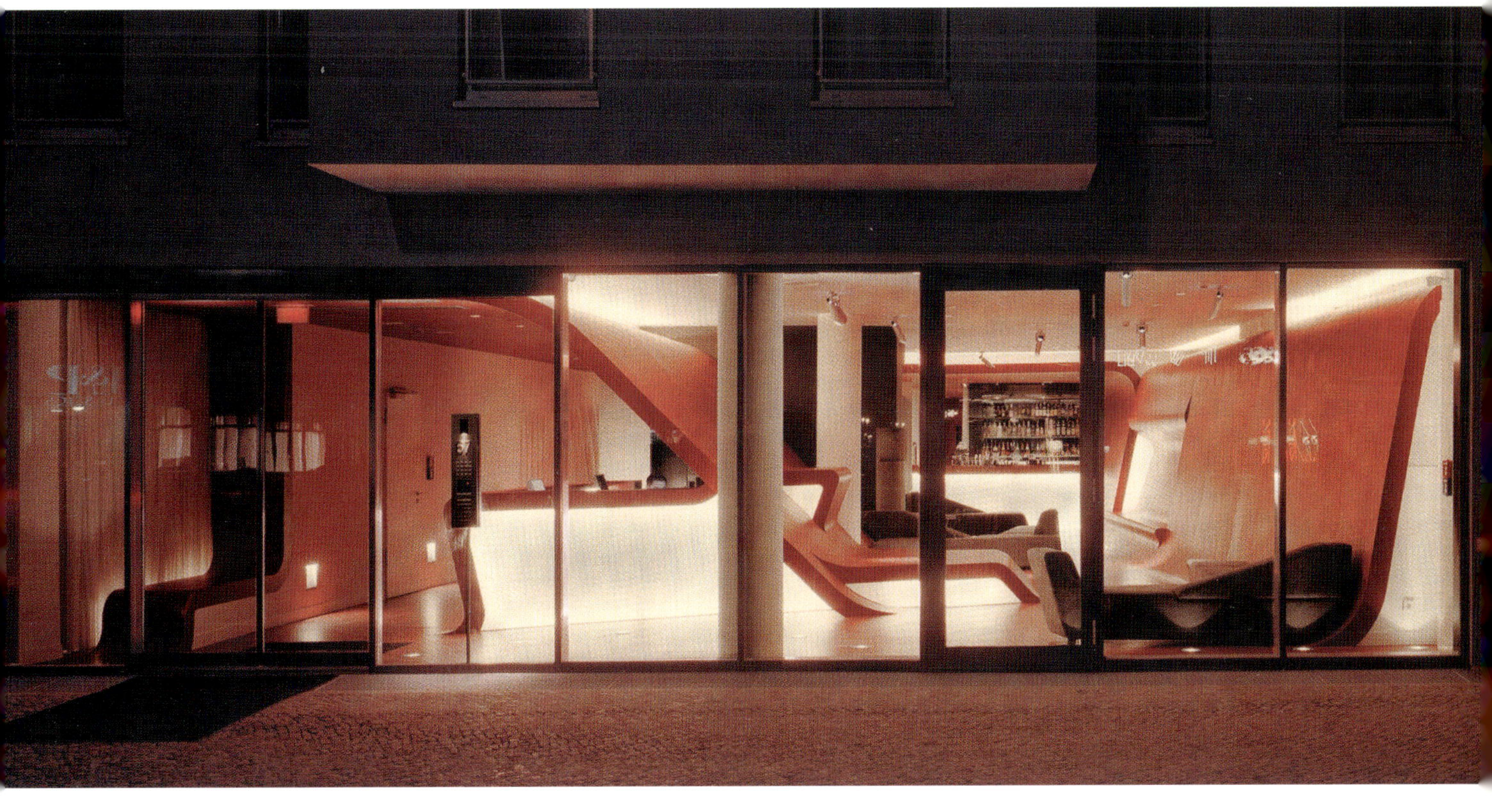

Fehn Sverre

Glacier Museum, Fjærland,
Norway, 1989–1991

Sverre Fehn (born in 1924) studied at the School of Architecture in Oslo, where he also lectured as a professor from 1971 to 1995. A trip to Morocco in 1952/1953 left a great impression on him and influenced his further development. Following this, he worked with Jean Prouvé in Paris. Fehn, who was meanwhile running a practice of his own in Oslo, collaborated with Geir Grung in 1955 in designing the widely admired Økern Home for the Elderly in Oslo. Fehn and Grung, together with a group of seven other young architects and Arne Korsmo, founded PAGON, the "Progressive Architects' Group, Oslo, Norway". Fehn gained international recognition with his design for the Nordic Pavilion at the Venice Biennale (1958–1962). This exhibition building illustrates many aspects of his architectural language: his materials of choice are largely restricted to concrete, wood, steel and glass. The roof construction featuring slender concrete ribs, measuring 6 centimeters by 1 meter in height, are unusual, giving the impression of lightness despite its material composition; the concrete is a mixture of white sand and pulverized marble. The sunlight is diffused, yet shadow effects are still created – a construction of "Scandinavian" light, which creates an appropriate background atmosphere for the works of art. Another of Fehn's major projects in the seventies was the Hedmark Museum in Hamar, Norway (1967–1979). In the nineties, he created the Villa Busk in Bamble, Norway (1987–1990) and the Glacier Museum in Fjærland, Norway (1989–1991), which illustrates his sensitivity in incorporating architecture into the landscape. This building, with its simple concrete walls, lies flat and unobtrusive against the mountain backdrop. The entrance area has a traditional gabled roof and simultaneously highlights the way into the interior, while the exterior stairways slope up to the roof terrace, as if pointing upwards to the mountain range. Fehn is thus making reference to the location and content of the museum. In 2000, the Ivar Aasen Centre for Language and Culture was realized in Ørsta, Norway. With its grass roof, it is set into the hillside and blends into the valley. His Museum of Architecture in Oslo (2007) will be housed in a classic-style bank building, with the addition of a separate exhibition pavilion.

"No matter how good an architect you are, if you have no chance of expressing your poetic idea in structures, you lack the very foundation of architecture. The structure is a language, a way of expressing yourself, and there should be a balance between thought and language."

Eisenman Peter

*"Architecture exists
only when it is
liberated from itself."*

Below: Greater Columbus
Convention Center, Ohio,
1989–1993

Opposite page: Wexner Center
for Visual Arts, Columbus,
Ohio, 1982–1989

Peter Eisenman (born in 1932) studied at Cornell University, Ithaca; the University of Columbia, New York; and Cambridge University, England. During 1957–1958 he was a member of The Architects' Collaborative (TAC), founded by Walter Gropius. In 1967, he founded the Institute for Architecture and Urban Studies in New York, of which he remained director until 1982. Eisenman designed a series of houses, which he numbered consecutively based on square ground plans and modified by distortions and deviations. Of these, the following designs were actually built: No. I, the Barenholtz Pavilion in Princeton, New Jersey (1967–1968); No. II, the Falk residence in Hardwick, Connecticut (1969–1970); No. III, the Miller residence in Lakeville, Connecticut (1969–1971); and No. VI, the Frank residence in Cornwall, Connecticut (1972–1975), with its non-functional, reversed, red staircase. During the eighties, Eisenman was responsible for a block of apartments at Checkpoint Charlie in Berlin (1981–1985) and the Wexner Center for the Visual Arts, part of Ohio State University in Columbus, Ohio (1982–1989). Contrasting with the white façades and dark windows that dominate the square part of the building are the variously shaped towers, grouped around the building. Along with Michael Graves, Charles Gwathmey, John Hejduk and Richard Meier, Peter Eisenman was one of the New York Five who adopted the architectural forms of the twenties. Gradually, however, he began to develop his own, esoteric theory of architecture, on the basis of which he was invited to exhibit at the "Deconstructivist Architecture" exhibition at the Museum of Modern Art (1988). One of Eisenman's most important projects was the Convention Center in Columbus, Ohio (1989–1993). His "Memorial to the Murdered Jews of Europe" in Berlin (2004–2005) is a place of meditation at the cutting edge of architecture and art, which manages to survive in a difficult environment.

Eames Charles ar

Below and opposite page:
Case Study House No. 8, Pacific
Palisades, California, 1945–1949

After attending the Washington University School of Architecture in St. Louis from 1924 to 1926, Charles Eames (1907–1978) opened his first practice. From 1937 he studied and taught at the Cranbrook Academy of Art in Michigan, where he met Eero and Eliel Saarinen as well as his wife Ray Eames (1912–1988). In 1940, in collaboration with Eero Saarinen, he designed a prize-winning, shell-shaped chair of moulded plywood. Charles and Ray Eames, who married in 1941 and moved to southern California, continued to work in experimental design using plywood. Their most famous furniture designs include "Lounge Chair", created in 1956; consisting of a lounger and stool, it was constructed of moulded, layered plywood on a curved, stainless steel frame (1958). The same sort of highly innovative approach also characterizes their architectural projects: their first steel frame construction design was for their own house, Case Study House No. 8 in Pacific Palisades, California (1945–1949), as part of the Case Study House Program. The building consists of two two-story cubes connected by a court-yard. The elevations are comprised of colored plates and glazed panels of varying sizes. By way of contrast, Case Study House No. 9 in Pacific Palisades, California, constructed by Eames and Eero Saarinen (1945–1949), is not an exposed structure but encased in wood cladding. Eames is, however, more famous for his film "Powers of Ten" (1977), in which he addresses the relative size of things in the universe.

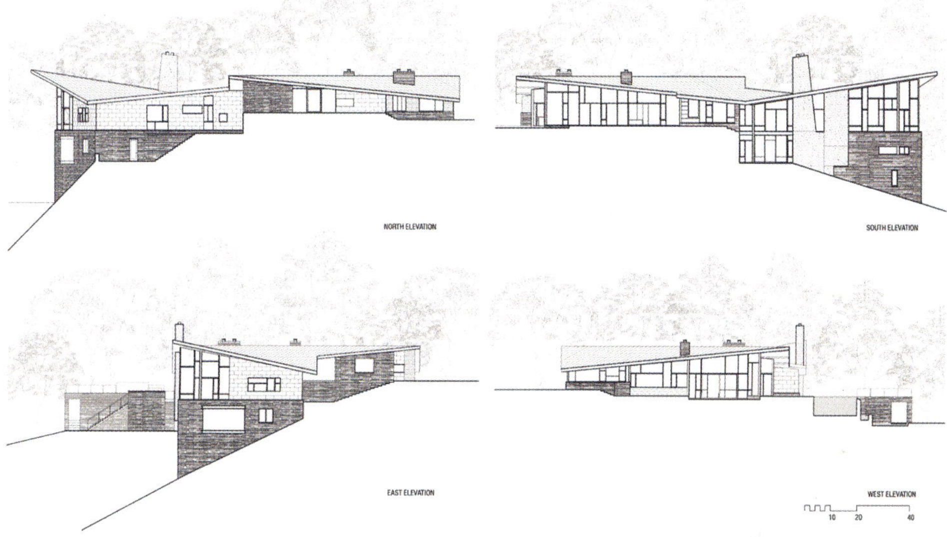
NORTH ELEVATION
SOUTH ELEVATION
EAST ELEVATION
WEST ELEVATION
10 20 40